LET'S DISCOVER

SPACE

igloo

This edition published in 2009
by Igloo Books Ltd
Cottage Farm
Sywell
NN6 0BJ

www.igloo-books.com

L006 1109

10 9 8 7 6 5 4 3 2 1

ISBN: 978 1 84852 848 2

Printed and manufactured in China

Written by: Dennis Ashton

Images provided by: NASA/ESO/ESA/SOHO/Paul Sutherland/
Shutterstock.com/istock.com/Clipart.com
Star maps created in conjunction with Stellarium software
(www.stellarium.org)

Contents

Please note:

Unless referred to, the references made to the positioning of the elements in the night sky are stated from a Nothern Hemisphere point of view. If you live in the Southern Hemisphere i.e. Australia, then simply reverse what is said i.e. Summer is your Winter.

Important!

Where you see this symbol in the book, please read. It gives information about health and safety.

Welcome to

Did you know that Jupiter is so large that the Earth could fit inside it 1,300 times? Or that Saturn could actually float on water if you could find an ocean large enough? And did you realize that the Sun will one day disappear from our skies?

Jupiter, Saturn and the Sun are among the many amazing objects that exist in our universe – scientists believe this vast space has been around for nearly 14 billion years and is about 90 billion light years across. Think about it: just one light year is over 5.6 trillion miles/nine trillion kilometers.

To make things even more incredible, it is believed that if you tried to cross the universe, you would end up where you began in the same way you would if you traveled round Earth.

The universe has been expanding since its creation. The young universe was made from hydrogen gas, with a touch of helium. This hydrogen made galaxies and the galaxies, in turn, made stars. The first stars spread dust and gas into space to make more stars and planets.

These galaxies and stars are scattered throughout the universe. Inside, stars and planets continue to be born. At the same time, old stars die, shrinking into white dwarfs, and some stars explode to make black holes – the scariest objects in the universe because no one is quite sure what would happen to you if you flew into one!

the Universe

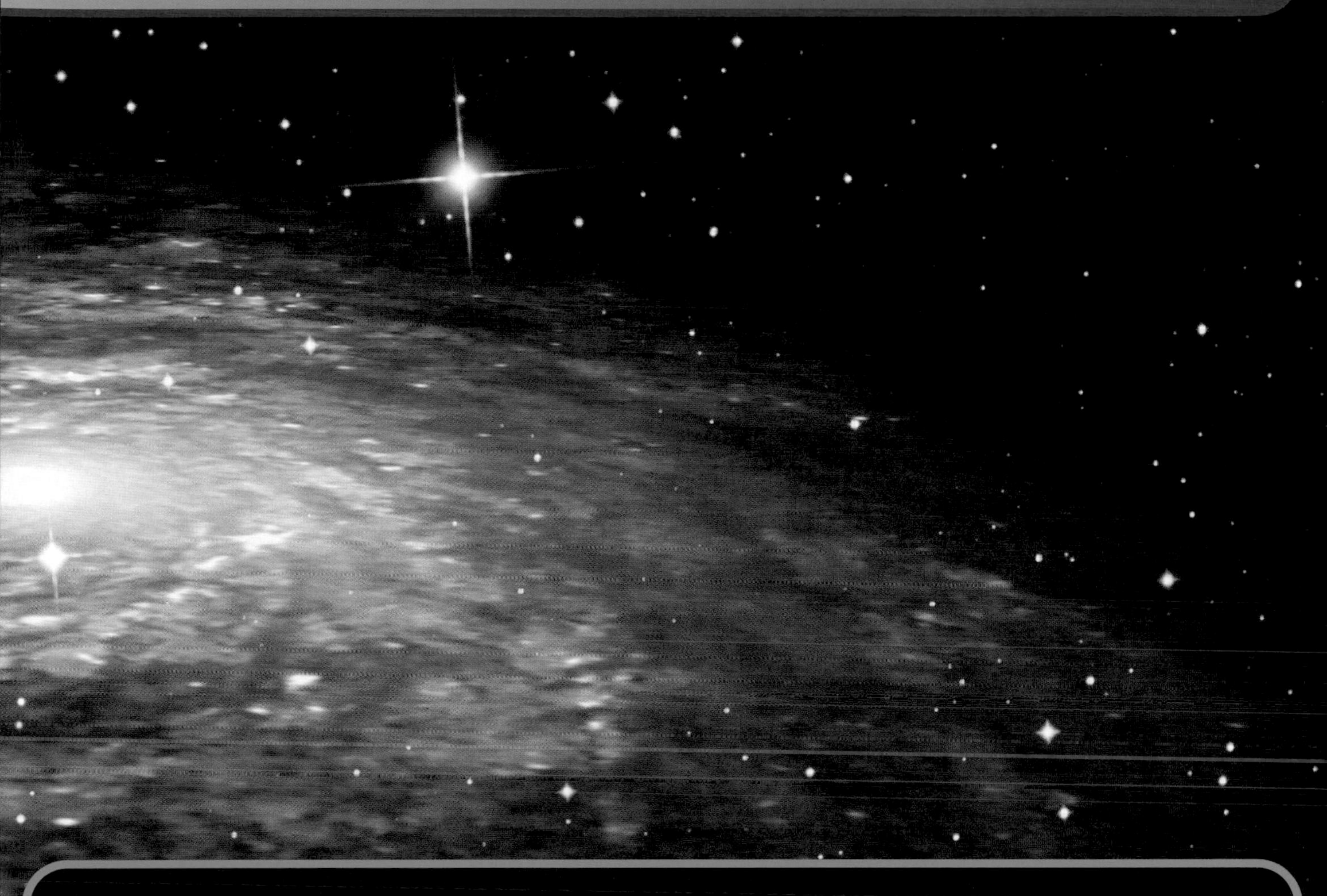

Our own star, the Sun, was made about five billion years ago, and the planets in our Solar System came along half a billion years later.

From our tiny planet Earth, we have sent spacecraft to visit other worlds and send back fantastic pictures. We have found that other stars have planets too.

We know all this thanks to astronomers. They are scientists who discover all this exciting information by looking at the sky.

With this book you can learn about all the great objects in the universe. The pages are packed with detail on stars, planets, nebulae and galaxies. Once you know what's out there in space, you can then take a look for yourself.

It will show you what you can see using just your eyes, and also tells you the best things to find in the night sky using a pair of binoculars or a telescope. There are star maps and sky close-ups to help you learn the star signs.

You will find out all about the amazing universe and explore its wonders – and very soon you will become an astronomer too!

Day & Night
Our Spinning Earth

Two thousand years ago, people were certain that the Sun moved across the sky. They saw the Sun rise in the morning, move slowly across the sky all day, and set at night. It was so obvious – and it was completely wrong!

We now know that it is not the Sun that is moving; it is the Earth that spins round every day and makes the Sun seem to move. It's like sitting on a carousel – you can see people moving past as the carousel spins. But of course, they are standing still as it's you and the carousel that are moving.

Planet Earth spins round once every 24 hours. The Sun lights the sky when we are facing it. When the Sun disappears, the sky goes dark and we can see

Precise Day

Clocks on Earth have a 24-hour day. In fact, the Earth rotates in 23 hours, 56 minutes, four seconds, but this is so close to 24 hours that it makes no real difference. It does mean that stars rise earlier by about four minutes each night. Over a week, stars rise earlier by about a half-hour. In a month it adds up to two hours.

Days on Other Planets

All the other planets have day and night, but they spin round at different speeds to our Earth.
The planet with the shortest day is Jupiter. It spins around in about ten hours, so Jupiter's day is only ten hours. There is a five-hour day and a five-hour night.
The longest day is on Venus, which rotates very slowly and takes 243 days to turn round once. So a day on Venus lasts 243 Earth days! The planet most like Earth is Mars. A day there lasts 24 hours, 38 minutes.

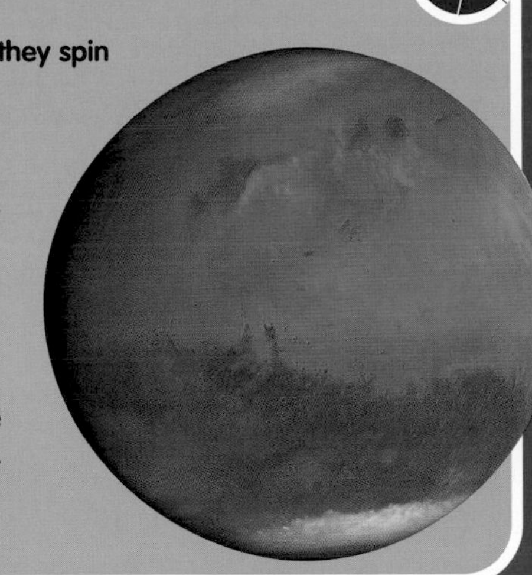

Planet Earth Statistics

Size (diameter):	**7,926 miles/12,756 km**
Distance from Sun:	**93 million miles/150 million km**
Day (rotation):	**23 hours, 56 minutes, 4 seconds**
Year (time to orbit Sun):	**365.25 days**
Temperature:	**71.6°F/22°C average**
Moons:	**1**

the stars. The Earth turns round from west to east so, looking out from our spinning Earth, the Sun rises in the east and sets in the west.

While we are in daylight, it's night-time for people on the other side of the world. There are different time zones around the Earth, so when it's one o'clock in the afternoon in London, it's midnight in Sydney, Australia.

At night, when the Sun has set, we can see the stars, but actually the stars are in the sky all the time. We can't see them during daylight because the Sun is so bright. As the Earth turns, the stars rise in the east, move slowly across the sky and set in the west. We see different stars in the sky before dawn than those we see after sunset.

There is just one star that does not move across the sky. It's the North Star, which is directly above the North Pole. The Earth spins around on its axis – which is a line through the Earth from North Pole to South Pole. So the star above the North Pole stays still and all the other stars seem to move slowly around it. The same would be true for the South Pole, but there is no bright star over this pole.

Years & Seasons

Spaceship Earth

We all know the date of our birthday. It comes around every year, every 365 days. We measure our age and history in years. A year is a very good way of measuring long periods of time – and it's our spaceship Earth that gives us this useful year.

Our Earth is a planet and, like other planets, it is in orbit around the Sun. One orbit takes a year, so your age tells you how many times you have been round the Sun.

The Earth does not stand up straight. Its axis, an imaginary line that runs through the Earth from the North Pole to the South Pole, it tilted over by 23.5 degrees. The tilt of the Earth causes the seasons to occur.

Summer

The North Pole is leaning towards the Sun, so in the northern half of the world there are long days and short nights. The Sun is high in the sky and gives hot days – summer in the northern hemisphere. The North Pole is in sunlight all the time, so it's daylight 24 hours each day.

The South Pole is facing away from the Sun, so it always stays dark there. In the southern half of the world there are short days and long nights – it's winter in the southern hemisphere.

Winter

Six months after the northern summer, the Earth is on the other side of the Sun. The North Pole is still leaning the same way, but now it's pointing away from the Sun and is in constant darkness. In the northern half of the world, there are short days and long nights. The Sun is low in the sky and gives less heat, so the days are

Speedy Planet

Our Earth really does travel through space like a spaceship. It orbits the Sun at 67,000 miles per hour/108,000 kilometers per hour. Astronauts take three days to reach the Moon but the Earth travels the same distance in only four hours. Spaceship Earth travels over 574 million miles/ 924 million kilometers every year.

Spring and Fall/Autumn

When the Earth is halfway between summer and winter, the poles are neither facing the Sun nor leaning away from it. Days and nights are the same length. The days are warm but neither very hot nor really cold.

cold. It's winter in the northern hemisphere.

In the southern halt of the world, the South Pole is facing the Sun, in constant daylight. The southern hemisphere has long days and short nights, so it's summer in the south.

Seasons do not only affect the way we dress. Plants adjust to the seasons by growing in spring, ready for full bloom in summer, then making seeds in fall/autumn to last through the cold winter. Some animals feed in summer and fall/ autumn, hibernate through the winter cold and then come back to life in spring. The movement of spaceship Earth affects everything on board.

Leap Years

If we measure in days, our Earth takes 365.25 days to orbit the Sun, so a year is 365.25 days. However, our calendar does not have quarter days, so every four years a whole day is added to take up these quarters, which makes a Leap Year. The date for the extra day is February 29.

Orbit

The Earth's orbit is not a perfect circle. We travel closer to the Sun on one side of our orbit and further away on the other side, but the difference is very small and has no effect on Earth. If you live in the northern hemisphere, the Sun is actually closer in the winter than in summer.

Light Pollution

Our View of the Sky

Two thousand years ago, people looked at the sky and saw the stars – and they knew the stars better than we do, with all our modern technology.

Ancient people had no 21st century technology, but their night sky was dark. They had no street lights or security lights, no shopping malls, offices or highways. They could see the stars in a way we never will.

In our cities we can hardly see any stars. The night sky is never completely dark because of all the lights we use. It's spoiled by light pollution.

We need lights to see at night and feel safe, but they should shine down onto the ground, not into the

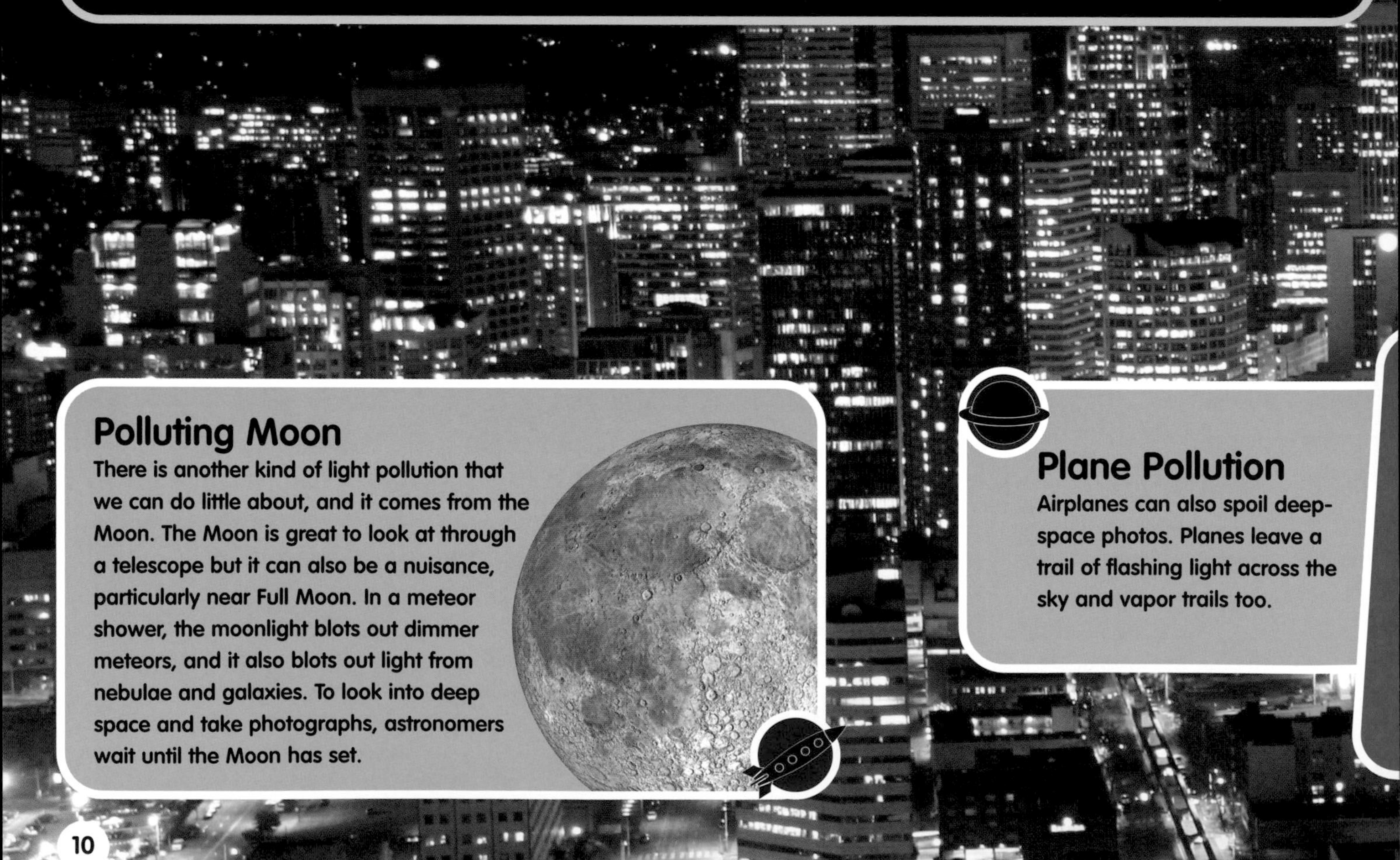

Polluting Moon

There is another kind of light pollution that we can do little about, and it comes from the Moon. The Moon is great to look at through a telescope but it can also be a nuisance, particularly near Full Moon. In a meteor shower, the moonlight blots out dimmer meteors, and it also blots out light from nebulae and galaxies. To look into deep space and take photographs, astronomers wait until the Moon has set.

Plane Pollution

Airplanes can also spoil deep-space photos. Planes leave a trail of flashing light across the sky and vapor trails too.

sky. Light in the sky not only spoils the sky but it's a waste of energy and money. And the wasted energy adds to global warming, too.

But there are answers to this problem. Shields on street lights help to cut down light pollution, and new types of low-energy lamps give better lighting without shining into the sky. They save money, they save energy and we can see the stars better, so everybody wins. Many cities are now using better lights, and astronomers are leading the way.

Light pollution is annoying, but the weather has the biggest effect on our view of the sky. Clouds are the astronomer's enemy. Light from stars can travel vast distances through space but can be stopped at the last second by clouds blotting out the sky making the stars disappear from sight.

We can't blow the clouds away, so an astronomer must be patient and hope for the sky to clear – or maybe go out to observe on another night. If you are planning a night of observation, keep an eye on the weather forecasts.

Light Pollution

The photo on the left shows the sky seen through city lights. The photo on the right shows the same stars from outside the city. Try counting the stars in each photo to see how light pollution affects the night sky.

The Moon

A world without air

After the Sun, our Moon is the biggest, brightest object in the sky. It is 25,000 times brighter than the brightest star, so bright that it can be seen in daylight. Yet the Moon does not make any light itself – it just reflects light from the Sun.

Moon Statistics

Size (diameter):	**2,156 miles/3,470 km**
Distance from Earth:	**238,600 miles/384,000 km**
Day (rotation):	**27 days, 8 hours**
Time to orbit Earth:	**27 days, 8 hours**
Full Moon to Full Moon:	**29 days, 12 hours**
Temperature:	**Day 225°F/107°C**
	Night -243°F/-153°C

Two thousand years ago, some scientists thought there were seas on the Moon. They imagined that dark patches were reflections of oceans on Earth. They called the dark patches 'Mare' (pronounced Mar-ray), which is the Latin word for sea. In fact, there is no water on the Moon. The Maria (plural of Mare) were made four billion years ago, when huge space rocks crashed into the Moon. The impacts made huge holes, many miles across. Lava flowed up from inside the Moon and filled the holes, then cooled and dried. So the Maria are not seas, they are lava plains.

The Moon is covered in craters made by the impact of meteorites. There are 1,545 craters with names and many smaller ones that have no names. Some craters have bright streaks, called rays, coming from them and spreading out over the lava plains. They were made by material blown out when the meteorites crashed into the surface.

Many craters have hills in the middle. This is where a space rock hit the Moon and blasted out the crater. The meteorite was blown to bits, but the impact crushed the ground below. The ground rebounded slowly to make a hill in the middle of the crater.

The edges of the Mare were

Moon Landing

Twelve American astronauts set foot on the Moon between 1969 and 1972.

Far Side

The same side of the Moon faces Earth all the time, so we can never see the far side of the Moon from this planet. But spacecraft have flown around the Moon and taken pictures of the far side, and it is quite different to the side facing Earth. It has hardly any lava plains but many more craters.

raised into mountains, which are lighter in color than the lava plains. Long valleys cross the Moon, sometimes cutting through the mountains. Some valleys are actually chains of craters.

The Moon is a world without air. Its gravity is too weak to hold an atmosphere around it and, if you stood on the Moon, the sky would be pitch black, even in daytime. There is no air to scatter the sunlight and make the sky blue. With no air, temperatures climb high in daytime to over 212°F/100°C and drop at night to minus 243°F/minus 153°C. Without air there can be no water and without water there can be no life, so the Moon is a lifeless planet. Some scientists believe there may be ice in cold craters near the Moon's poles. If there is any ice there, it could have come from crashing comets.

When people began to launch rockets into space, the Moon became the number one target. In 1959, a Russian Luna space probe crashed into the Moon. Luna 9 made the first soft landing in 1966. Three years later, American Neil Armstrong became the first human to stand on the Moon.

When you look at the Moon, you are looking at the only other world so far visited by humans.

The Moon

A different look every night

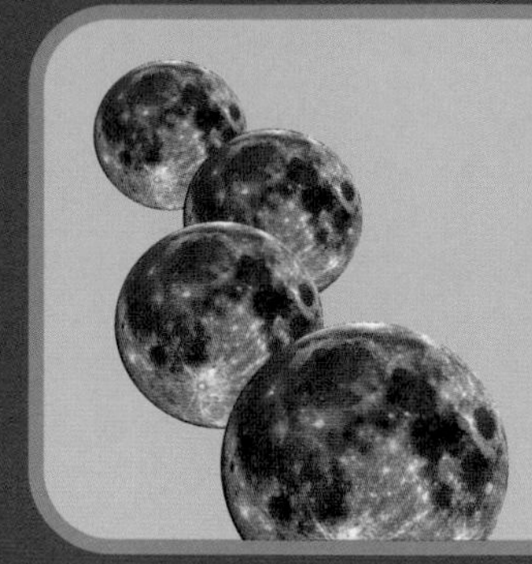

Same Face, Different Phase

The Moon spins round once every 27 days, always keeping the same side facing the Earth. From Earth, we can only see one side of the Moon – we cannot see the far side.

The Moon looks different every night. It changes, slowly and regularly, because it is in orbit around the Earth and reflects light from the Sun. The changing appearance produces the phases of the Moon.

The Moon orbits the Earth every 27 days, so let's see how the Moon looks at different places in its orbit in one month. In our picture, the Sun is shining from the right. The Earth is at the middle and the Moon goes round the Earth.

Notice that the Sun always lights up half of the Moon. That does not change. From Earth we see the Moon in a different place. We see only some of the light that bounces off the Moon.

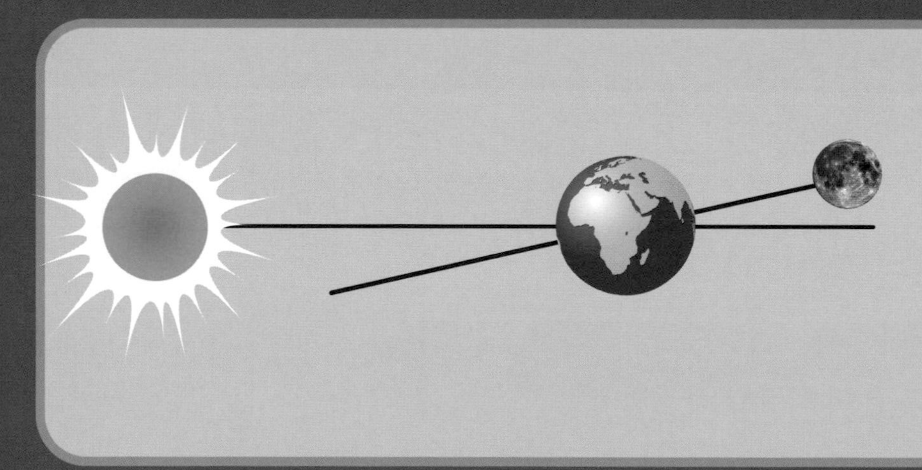

Tilted Orbit

The Moon's orbit is slightly tilted over, at an angle to the Earth's orbit. At New Moon, the Moon is between the Earth and the Sun. We could expect an eclipse but the Moon is just above or below the Sun. This means that we do not have an eclipse of the Sun each month. At Full Moon, something similar happens. The Moon is above or below the Earth's shadow, so there is no eclipse. Eclipses are explained in more detail on pages 20–23.

1 New Moon
The Moon is between the Earth and the Sun. The sunlit side of the Moon faces away from us. The dark side faces us – and the Moon is hidden by the glare of the Sun. We cannot see the Moon at all.

2 Waxing Crescent
The Moon is no longer between the Earth and the Sun. We begin to see part of the sunlit side as a thin slice. This is a crescent Moon, and waxing means that it will become bigger the following night.

3 First Quarter – Half Moon
The Moon is one quarter of its way around the Earth. It is now about a week since new Moon. Looking from the Earth, we can see half of the sunlit side, so we see a Half Moon.

4 Waxing Gibbous
Here, from the Earth, we can see most of the Moon lit up. If more than half is lit up, it is called a Gibbous Moon.

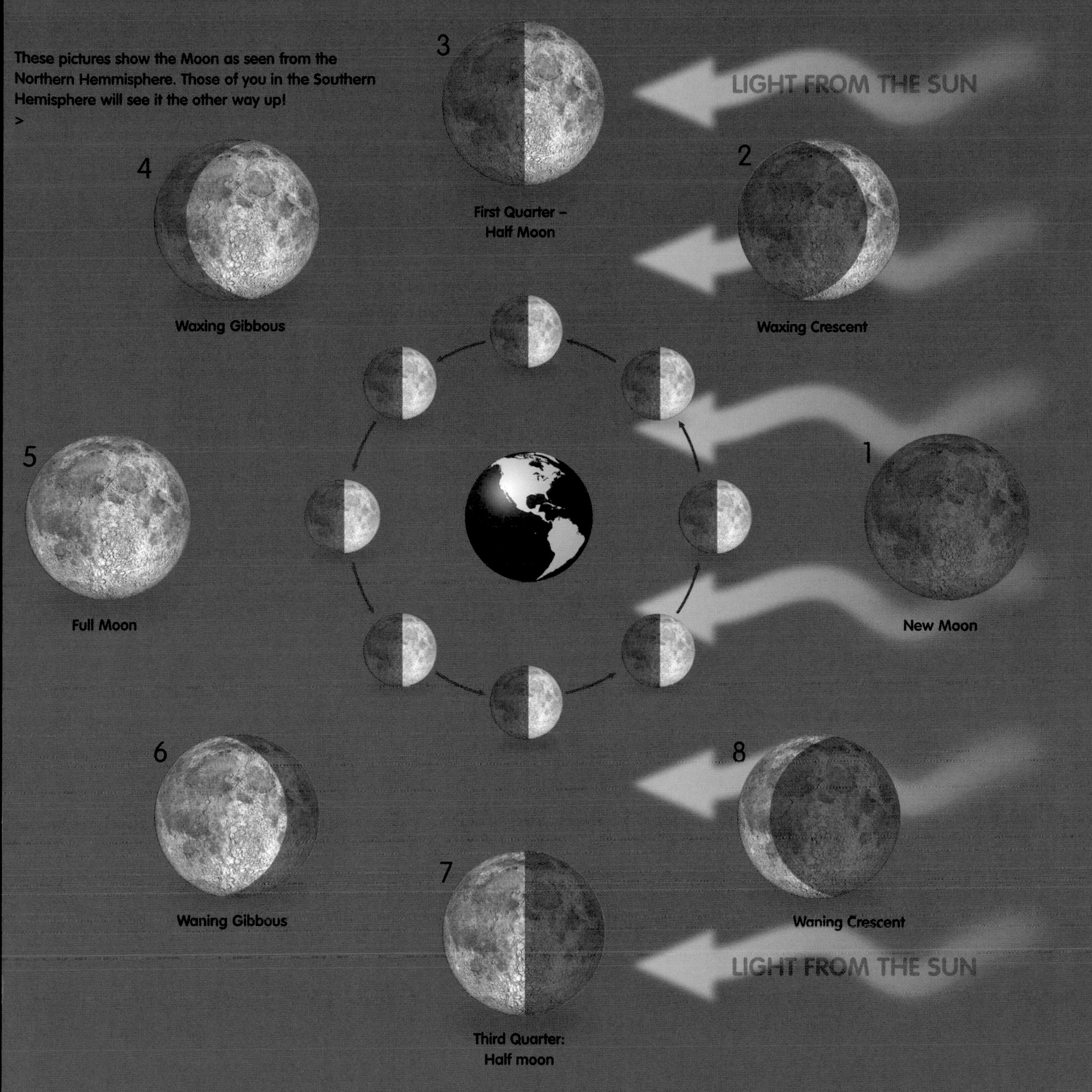

5 Full Moon

It is two weeks after New Moon. The Moon has arrived on the opposite side of the Earth to the Sun. Now we can see the whole of the sunlit side. This is Full Moon.

6 Waning Gibbous

The Moon's orbit carries it back towards the Sun. Here we see most of the sunlit side. It is a Gibbous Moon and the sunlit part will become smaller and smaller. This is a Waning Moon.

7 Third Quarter: Half Moon

The Moon is three-quarters of its way around the Earth, and we can see half of the sunlit side. This is Half Moon. At First Quarter, one side of the Moon was sunlit. Now we see the other side lit up.

8 Waning Crescent

The Moon has almost finished one orbit. We can see only a thin slice of the sunlit side. This is a crescent Moon, and we can see it in the morning, before dawn. It is waning and soon it will be New Moon again.

The Sun

Our Local Star

We think our Earth is a pretty important planet, but in fact we live on a piece of dirt. Our Solar System is really just the Sun with a few pieces left over, and we live on one of those pieces.

Sun Statistics

Size (diameter):	**870,000 miles/1,400,000 km**
Surface temperature:	**10,500°F/5,800°C**
Core temperature:	**23,400,000°F/13,000,000°C**
Rotation (spin time):	**27 days**
Distance from Earth:	**93,000,000 miles/150,000,000 km**
Time for light to reach Earth:	**8 minutes, 20 seconds**
Star type:	**G2**

The Sun is 99.8% of the Solar System, so it's very big compared to planets. It's a star, but it looks bigger and brighter than all the other stars because it's close to us. Light from other stars take years to reach us – they are light years away – while light from the Sun takes only 8.3 minutes to reach the Earth.

Although it is an ordinary star, the Sun is special to us because it's the only star we can study really well. A lot of what we know about other stars comes from what we know about the Sun. So what do we know?

The Sun is at the center of the Solar System. All the planets are in orbit around it. The Sun holds planets in orbit with its powerful gravity.

The Sun is made of hydrogen gas mixed with helium – 75% hydrogen and 25% helium. The Sun is like a ball of burning gas, and its surface is about 10,500°F/5,800°C – hot enough to melt any metal.

SOHO

The Solar and Heliospheric Observatory (SOHO) is a space observatory in orbit that watches the Sun all the time, taking pictures in different kinds of light. Here is a SOHO image of the active Sun.

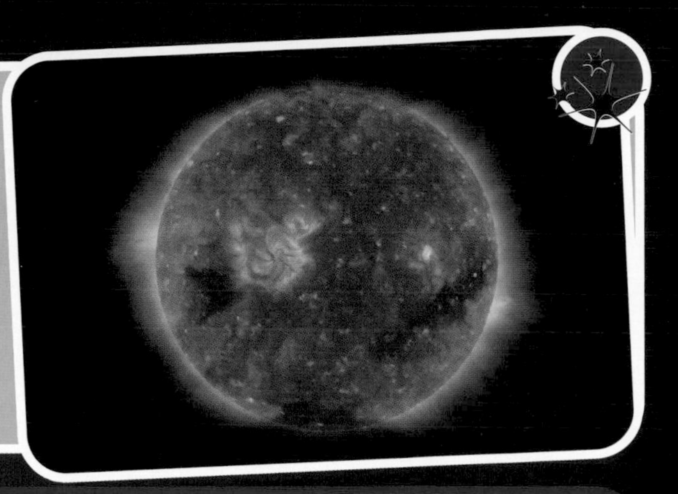

The surface of the Sun looks bright yellow, and on the gassy surface we often see dark spots called sunspots. These are little patches of cooler gas, about 7,000°F/ 4,000°C. Sunspots, some of which are bigger than the Earth, are controlled by magnetic fields on the Sun. They move slowly across the Sun, which proves that the Sun spins round, like planets. The Sun takes about 27 days to turn round once.

Sometimes there are lots of sunspots, at other times almost none. The number rises and falls every 11 years, and the next high sunspot numbers will come around in 2010 and 2011.

Sunspots are active areas, from which big explosions called solar flares can occur. Flares blow material into space that, if it blasts our Earth, would damage satellites and cause a natural light show in the night sky. This effect is known as the Northern Lights if you are in the northern hemisphere, or the Southern Lights if you are in the southern hemisphere. Northern Lights are explained in more detail on pages 40–41.

At times, huge loops of fire, called prominences, leap and fall back to the Sun. They are pillars of gas that flow with the magnetism of the Sun, which is a bubbling, burning sphere of gas.

Danger!

You must never try to look at the Sun with binoculars or a telescope, because the heat and light from the Sun will burn the inside of your eyes. It will damage your eyes for the rest of your life, and could make you blind.

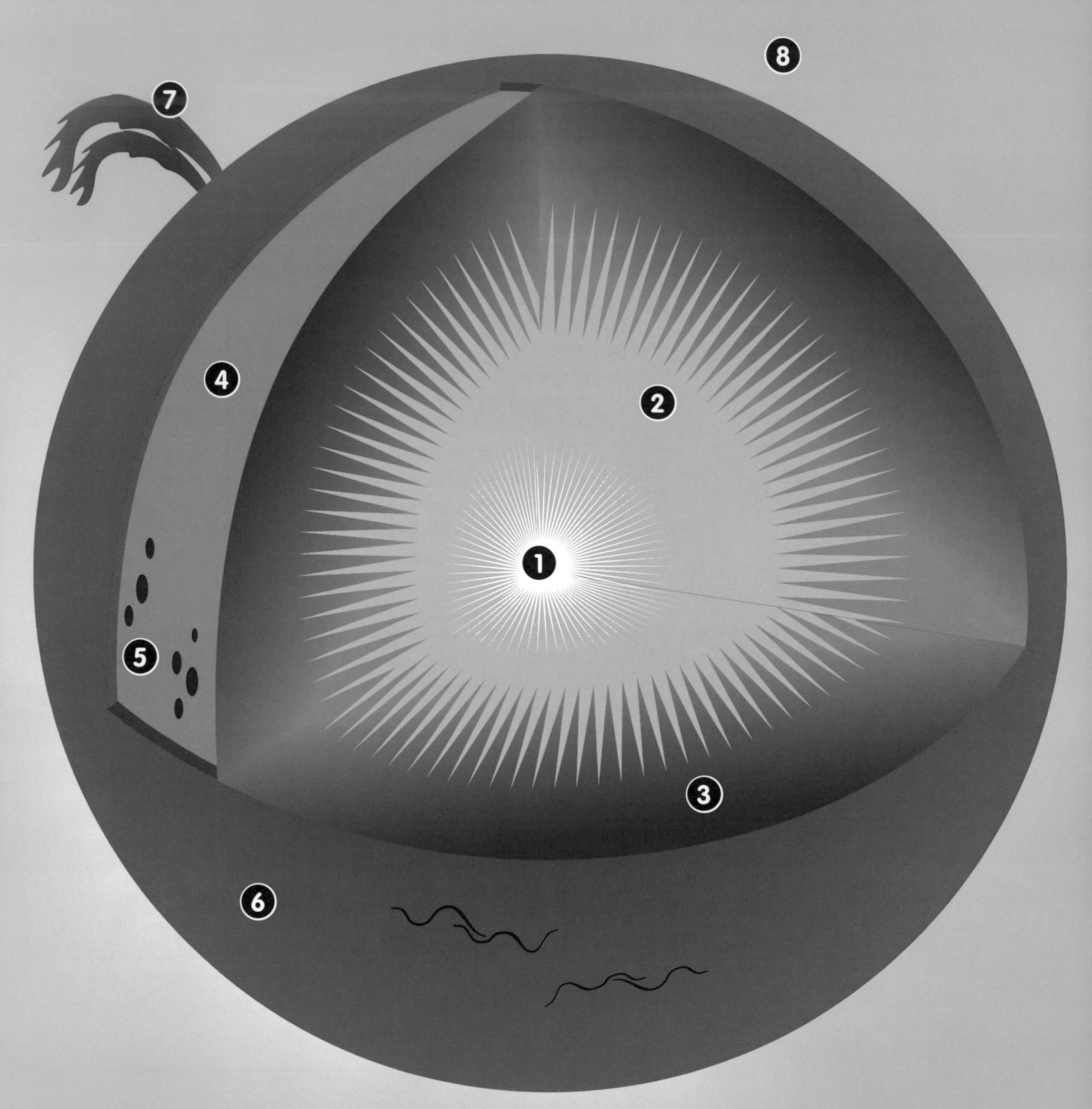

1 Core
The middle of the Sun is its powerhouse. Hydrogen gas is turned into helium in a non-stop explosion, and the energy from the explosion makes all the heat and light of the Sun. The temperature in the core is a staggeringly fiery 13 million degrees Celsius.

2 Radiative Zone
Energy from the core spreads into the gas around the core. The energy moves like heat and light coming from an electric fire. This is called energy radiation. But the energy moves slowly – heat can take a million years to move through this part of the Sun!

3 Convective Zone
Hot gas rises towards the top of the Sun. It takes energy with it. The gas bubbles to the surface of the Sun, loses its energy at the surface and cools down. The cool gas sinks back into the Sun and, when it's reheated, bubbles back up again.

4 Photosphere
This is the Sun's surface. Energy from inside bubbles up to the surface. The photosphere is covered in huge bubbles called granules. The temperature here is 10,500°F/5,800°C. The photosphere gives out heat and light. Some of this energy reaches Earth and gives us life.

The Sun

Inside Our Local Star

That's Big

The Sun is big! It is about 106 times wider than the Earth, and about 1,300,000 Earths would fit inside the Sun.

We cannot see inside the Sun, but astronomers have worked out what is happening inside our nearest star. And they've done it by collecting observations and using powerful ideas.

Their work shows that the Sun is three-quarters hydrogen gas and one-quarter helium.

Some astronomers observe the Sun during eclipses. Other astronomers listen to the Sun, for sound waves that give clues to the different layers inside our local star.

The most important idea came from Albert Einstein, who explained how the Sun could produce lots of energy from its hydrogen gas.

All these observations and ideas have been put together, and we now know that there are different layers of gas where different events are happening.

Gas Burner

The core of the Sun uses up four million tons of hydrogen gas every second, and it has been doing this for five billion years. Yet the Sun still has enough gas to last for another five billion years.

Galactic Year

The Sun and Solar System are about 26,000 light years from the center of our galaxy, the Milky Way. The Milky Way turns like a slowly spinning wheel and the Sun takes about 225 million years to turn with our galaxy. This is called a galactic year. The Sun is about 22 galactic years old.

5 Sunspot
Sunspots are cooler patches in the photosphere with a temperature of around 7,000°F/4,000°C. The gas is cooler because the Sun's magnetism stops heat coming from inside. Sunspots – some of which are bigger than our Earth – are controlled by magnetism.

6 Chromosphere
Above the photosphere is a thin layer of gas, which can be seen in an eclipse. It glows red and is called the chromosphere, as 'chromo' means 'color'. Lots of tiny flames called spicules sprout from this layer, looking like blades of burning grass blowing around in the wind.

7 Prominence
Sometimes giant clouds of flame called prominences leap from the Sun. A prominence can hold 100 billion tons of material and some grow to 125,000 miles/ 200,000 kilometers. Some make loops, like arches of fire, others erupt suddenly and blast material into space.

8 Corona
The corona is the outer layer of the Sun. It can be seen in a total eclipse of the Sun, when it looks like a ghostly blue-white halo around the Moon. The temperature of the corona is around one million degrees Celsius.

Eclipse of the Sun

Night-time in the Day

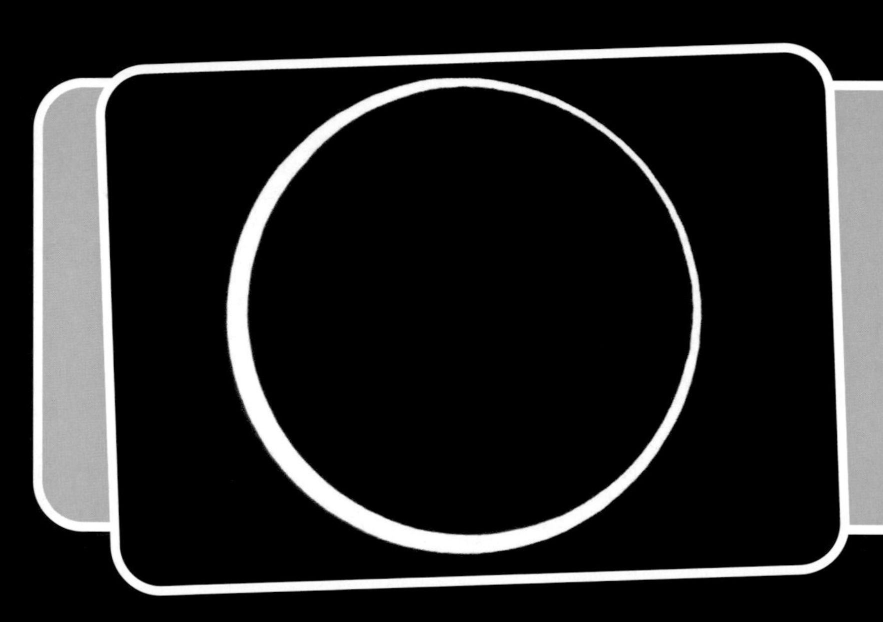

Annular Eclipse

The Moon's orbit is not a perfect circle, so sometimes it's a little closer to the Earth, sometimes further away. If an eclipse happens when the Moon is further away, it does not quite cover all the Sun. This is when we see an annular eclipse. The Sun makes a bright ring around the Moon.

You must try to see a total eclipse of the Sun as it's a dramatic sky event.

In a total eclipse, the Moon covers the Sun. The sky goes dark blue-black in daytime, you can see planets and bright stars, it gets cold and birds and animals go to sle[illegible] sunlight is blocked out apart from a ghostly white glow around the Moon. This g[illegible]he Sun's corona, and an eclipse is the only time you can see this outer layer. The eclipse only lasts a few minutes, but it's a few minutes you will never forget.

A total eclipse of the Sun happens when the Moon comes exactly between the Earth and the Sun, casting a shadow onto the Earth. It happens once or twice each year somewhere in the world. If you are outside the shadow, you may see a partial eclipse, when the Moon covers part of the Sun.

You will be in a part shadow called the penumbra. But the best place to be is in the full shadow, the umbra – then you can see the total eclipse. Because the Moon is so small, the full shadow is small and only a few places on Earth fall into it.

You have to be in exactly the right place at the right time, but you will have a choice of places to visit because the shadow moves across the Earth as the Earth turns.

Some Total Eclipses and Where to See Them

2008	2009	2017
August 1	July 22	August 21
China & Russia	India & China	Across the USA

Jupiter's Eclipses

The four big moons of Jupiter can eclipse the Sun. We sometimes see the shadow of a moon on Jupiter as a small dark circle. If you could stand inside the shadow on Jupiter, the moon would cover the far away Sun. Pluto's moon, Charon, can also eclipse the Sun.

Eclipse of the Moon

When the Moon goes Red

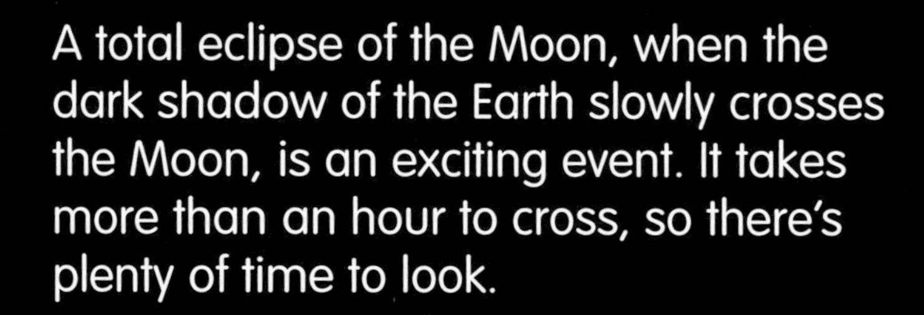

A total eclipse of the Moon, when the dark shadow of the Earth slowly crosses the Moon, is an exciting event. It takes more than an hour to cross, so there's plenty of time to look.

A lunar eclipse can only happen at Full Moon, when the Moon is on the opposite side of Earth to the Sun. At the start of an eclipse, the Earth's dark shadow (curved because our Earth is round) begins to creep slowly across the bright Full Moon. As the Moon moves across the sky, the shadow blots out more and more of the Moon and makes craters and mountains stand out clearly.

It takes about an hour for the shadow to cover the Moon completely. Then the eclipse becomes total. At total eclipse, the Moon is in the Earth's shadow, so you might expect it to be completely dark. After all, the Moon only shines by reflected sunlight and, with no sunlight, it should disappear into darkness. But something remarkable happens instead.

The Earth's air bends sunlight into the shadow and our air also turns this sunlight red – so weak red sunlight falls on the eclipsed Moon and it turns blood red! The color depends on how much dust is in the Earth's air: the more dust, the redder the eclipse.

Binoculars or a telescope will show the eclipse in close-up, and you might try to take photographs too. After an hour, the Earth's shadow slowly leaves the Moon and, sixty minutes later, the shadow has gone and the Full Moon returns.

It is quite safe to watch a lunar eclipse, for the Moon shines only by reflecting sunlight so there is no danger to your eyes. You do not need to wear eclipse glasses!

Some Moon Eclipses and Where to See Them

Most of them are just partial eclipses. But here are a few total eclipses coming soon.

2008
February 21
Times: 1.45am to 5.10am GMT
Total eclipse 3.00am to 3.50am GMT
North America, South America, Africa, Europe including Great Britain

2010
December 21
North America

2011
June 15
Africa, Asia, Australia

2015
September 28
Times: 1.08am to 4.28am GMT
Total eclipse 2.11am to 3.25am GMT
North America, western Europe including Great Britain

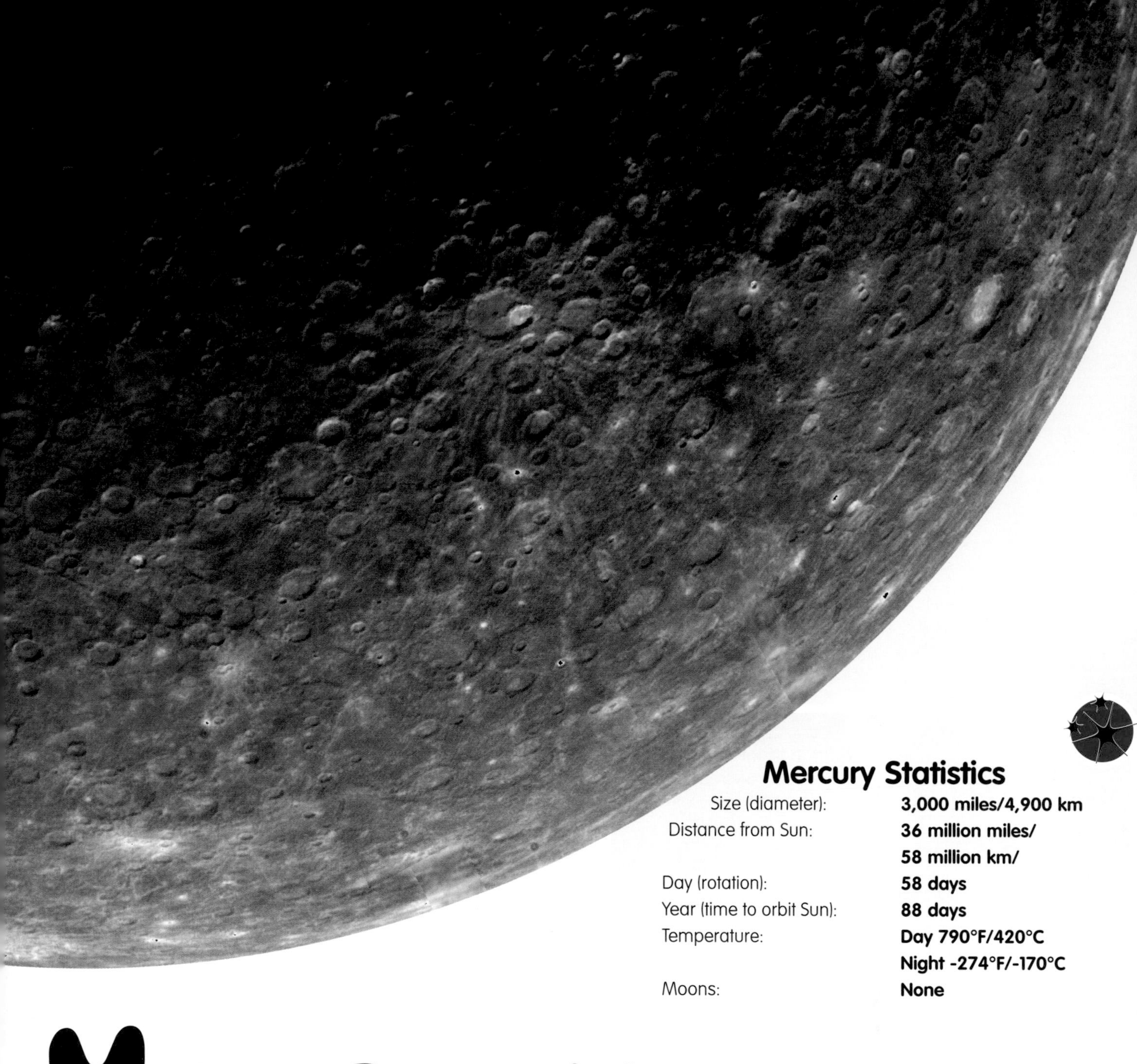

Mercury Statistics

Size (diameter):	**3,000 miles/4,900 km**
Distance from Sun:	**36 million miles/ 58 million km/**
Day (rotation):	**58 days**
Year (time to orbit Sun):	**88 days**
Temperature:	**Day 790°F/420°C**
	Night -274°F/-170°C
Moons:	**None**

Mercury

The Moon-like Planet

Mercury, which is the closest planet to the Sun, looks like the Moon. It is a small, rocky world and, like our Moon, it's covered in craters. Space rocks crashing into Mercury long ago made these craters.

In the daylight, close to the Sun, temperatures on Mercury climb to 790°F/420°C, twice as hot as an oven. In the dark of night, the temperature drops to -274°F/-170°C, much colder than a freezer. Mercury has no air or water, so there is no life on this little world.

Even the most powerful telescopes on Earth cannot tell us what Mercury is like. They just show that Mercury has phases like the Moon as it orbits the Sun. Most of our knowledge comes from Mariner 10, a robot spacecraft that flew close to the planet in 1974. Photos from Mariner 10 showed a planet scarred by thousands of craters, made by meteorites that crashed into the planet four billion years ago. The largest crater is the Caloris Basin, over 800 miles/1,300 kilometers wide. Some craters have lighter-colored rays spreading from them; this is material blasted out by the impacts.

Huge cliffs on Mercury rise two miles/three kilometers from the surface, probably made as the planet cooled and cracked long ago.

If we ever land on Mercury, we are sure to have a strange day. On Earth, a day lasts 24 hours, but on Mercury it lasts 176 Earth days. Mercury's orbit also takes it closer to the Sun, then further away, so the Sun would sometimes look big and sometimes small, depending on where the planet is in orbit.

Mercury Names

Craters on Mercury are named after famous artists, writers and musicians. Among them are Beethoven, Michelangelo, Shakespeare and van Gogh.

Mercury Messenger

No spacecraft has visited Mercury since 1974 and, to find out more about this little world, we need another robot to make a visit. A new probe Mercury Messenger will reach Mercury in 2008 and settle into orbit in 2011.

Venus
The Planet on Fire

In size, Venus is like a twin of our Earth. But it's far from being an identical twin, for Venus is a fiery furnace of a world.

Venus glows like a diamond in the sky, brighter than any star. In fact, it looks so beautiful that it was named after the Roman goddess of love. Venus glows brightly because it comes closer to the Earth than any other planet and is covered in cloud that reflects a lot of sunlight, and can sometimes be seen in daylight.

Venus Statistics

Size (diameter):	**7,521 miles/ 12,104 km**
Distance from Sun:	**67 million miles/ 108 million km**
Day (rotation):	**243 days**
Year (time to orbit Sun):	**225 days**
Temperature:	**860°F/460°C**
Moons:	**None**

Names on Venus

Venus is the only planet with a woman's name and its volcanoes, mountains, plains and craters are also named after women, either goddesses or famous women from history. Among the craters are Cleopatra, Mona Lisa, Curie and Nightingale. The exception is Mount Maxwell, named after the Scottish scientist James Clerk Maxwell.

The clouds always cover Venus completely, hiding the surface from us. Even the most powerful telescopes on Earth cannot see through the clouds.

In the past, scientists often tried to imagine what Venus could be like beneath those clouds. Some thought it was like Earth, with forests and animals, while others thought there were huge oceans of water. Some believed Venus was covered by hot desert. The truth was hidden until 1989.

In that year, a robot spacecraft called Magellan, with cameras that did not take ordinary pictures, arrived at Venus. Magellan had radar cameras that could see through the clouds and make a map of the surface. What the pictures showed was that, for humans,

The spaceprobe Venus Express over the planet's south pole

Strange Spin

Venus spins round on its axis in 243 days, so a day on this fiery planet lasts for 243 Earth days. And Venus orbits the Sun in 225 days, so a Venusian day lasts longer than its year! Not only does Venus spin more slowly than any other planet, it also spins round backwards compared to other planets. The Sun rises in the west and sets in the east. We don't know why Venus spins backwards.

Venus is a nightmare world. The clouds are made of sulphuric acid and the air is mostly carbon dioxide, a gas that would suffocate a human. The air traps the Sun's heat like a blanket, and global warming raises the temperature to about 860°F/460°C, all the time.

Venus is much hotter than an oven, and the air pressure is so heavy that it would crush a human as flat as a pancake.

An astronaut landing on Venus would be attacked by acid, suffocated, roasted and squashed! No life could survive on this fiery world.

It's also a world of volcanoes and lava flows, and some of the volcanoes may still be erupting now. Two huge continents rise above the volcanic plains. In the north, Ishtar Terra is the size of Australia. On Ishtar is the highest mountain, Mount Maxwell, which is higher than Mount Everest. In the south, Aphrodite Terra is the size of South America. Like other rocky planets, Venus has mountains, valleys and craters – and unusual sights too. There are flat volcanoes and ridges making patterns like spiderwebs.

In the 1970s, the Russian Venera spacecraft managed to land on Venus and sent back a few pictures of the hot, dry surface. But after a couple of hours, the signals stopped. The probes could not survive the hot, acidic, crushing atmosphere.

Mars Statistics

Size (diameter):	**4,225 miles/6,800 km**
Distance from Sun:	**142 million miles/228 million km**
Day (rotation):	**24 hours, 37 minutes**
Year (time to orbit Sun):	**687 days**
Temperature:	**+79°F/+26°C to -215°F/-137°C**
Moons:	**2**

Mars
The Red Planet

Mars is probably the most exciting planet in the Solar System. It's certainly the world most visited by robot spacecraft. Although it's smaller and cooler than Earth, Mars might just be warm enough for life.

Even using a big telescope, Mars can be an annoying planet. Its small size and our wobbly atmosphere make its markings difficult to see clearly. Some astronomers a hundred years ago thought they could see canals criss-crossing the red planet, and said intelligent Martians had made them. This belief inspired HG Wells to write his book 'War of the Worlds', in which Martians invade the Earth.

But when, in 1964, the robot spacecraft Mariner 4 flew past Mars, it destroyed all hopes of life. It showed the planet was a dry, desert world with no canals and no signs of life.

Since 1964, many more robots have visited Mars and, amazingly, our hopes of finding life there have risen again. Spacecraft have been sent into orbit around the planet and we now know more about that world than about any other planet except Earth.

From orbit, Mars is an orange-red planet, looking like a desert world. Sand gives the planet this color, and dark markings are areas of rock where the sand has blown away. At the North and South Poles, ice caps gleam snow-white.

Huge volcanoes sprout from the surface like giant pimples. One of them, Olympus Mons, is the biggest volcano in the Solar System. It is 342 miles/550 kilometers wide and 17 miles/27 kilometers high – three times higher than Earth's Mount Everest. If Olympus Mons were placed in Europe, it would cover most of Germany! All the volcanoes are now extinct, but when they erupted in the past they would have made Mars quite a different planet.

Water on Mars?

In 2005, NASA scientists saw a fresh white mark in a gully on Mars. It looked like a flow of water, so perhaps there is liquid water underground. Water is essential for life, so there is still hope of finding some kind of life on Mars.

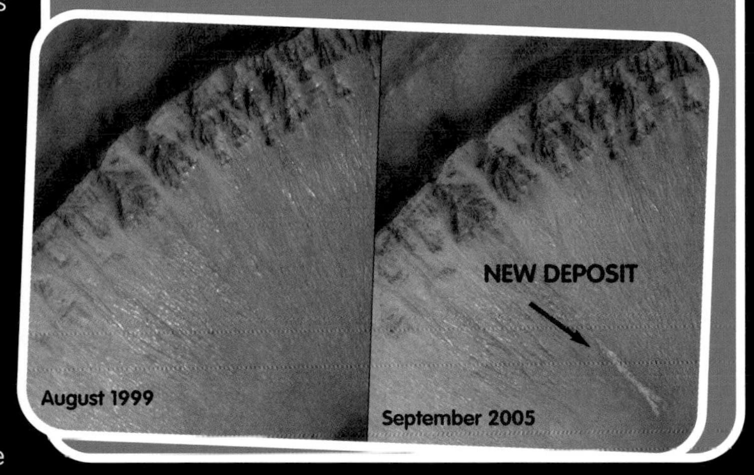

A huge gash, called Mariner Valley, cuts across one side of Mars. It is 2,800 miles/4,500 kilometers long, 373 miles/600 kilometers wide and 3.8 miles/six kilometers deep. If it were on Earth, Mariner Valley could stretch all the way across Europe.

Meteorite craters are found all over Mars, for the thin air gives little protection from incoming space rocks.

Close-up pictures taken from orbit show river valleys. They are dry now, but the valleys indicate that water did flow on Mars some time in the past. It seems that Mars was a warmer planet long ago, with water flowing in rivers and lakes. Perhaps in that water there was some kind of life.

Mars has two oddly shaped moons called Phobos and Deimos. Their names mean 'fear' and 'panic', so they fit well with the name Mars, who was the Roman god of war. The moons are probably asteroids that wandered too close to Mars and were captured by its gravity.

Living on Mars

When humans set up bases on Mars, they will feel at home in one way. Mars spins around every 24 hours, 37 minutes, so day and night on Mars are similar to those on Earth. The planet's axis is also tilted like Earth's, so it has seasons – but because Mars takes almost two years to orbit the Sun, the seasons are twice as long.

Jupiter
King of the Planets

Jupiter is truly a giant planet. All the other planets in the Solar System could fit inside Jupiter and there would still be room to spare, and it could swallow 1,300 Earths.

Jupiter is a different kind of planet to Earth, and is known as a gas giant. It's made of gas and has no solid surface, so no one will ever land there.

Although it is huge, Jupiter spins around amazingly quickly. A day lasts only ten hours, and this fast spin gives the planet its striped appearance. White ammonia and methane, and red sulphur chemicals are spun into colored bands around the planet, and beneath these clouds is a deep ocean of hydrogen gas.

Jupiter is so big that the cloud stripes can be seen through a telescope. Spacecraft have taken close-up pictures of the clouds, showing whirls of gas as the clouds swirl in powerful winds and storms.

The greatest storm is the Great Red Spot, a hurricane that has blown for over 300 years. Many more whirlwinds appear then disappear in the clouds. The latest was a red storm called Red Spot Junior.

Jupiter has 63 moons. Most of them, fairly small and irregular in shape, are thought to be asteroids captured by this huge world's mighty gravity.

Jupiter Statistics

Size (diameter):	**88,200 miles/142,000 km**
Distance from Sun:	**483 million miles/778 million km**
Day (rotation):	**9 hours, 50 minutes**
Year (time to orbit Sun):	**12 years**
Temperature:	**-238°F/-150°C**
Moons:	**63**

Inside Jupiter

At the heart of Jupiter there is probably a solid rocky core about the size of Earth. Above this core, the planet is made of hydrogen, which is so squashed that it becomes a liquid. Most of Jupiter is made of hydrogen liquid and only the top 620 miles/1,000 kilometers are hydrogen gas. Jupiter gives out more heat than it takes in from the Sun. Its core is hot, about 36,000°F/20,000°C. This heat is probably left over from when Jupiter was made, more than four billion years ago.

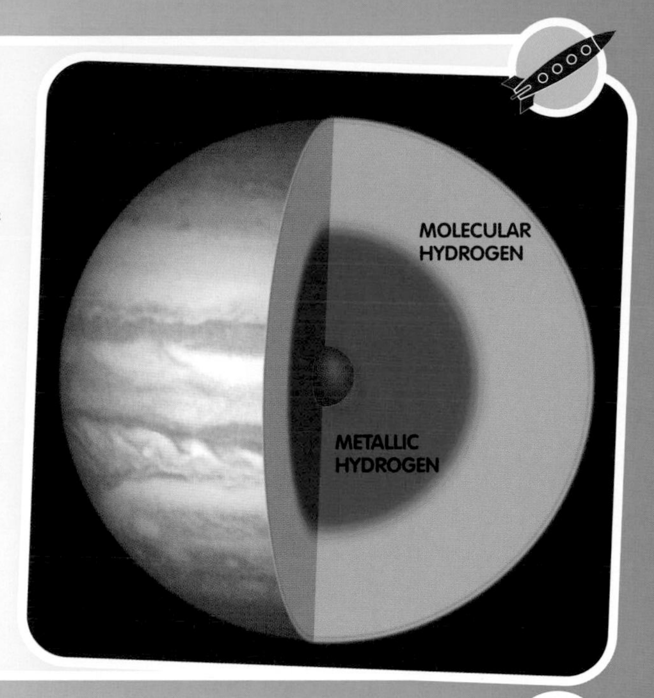

Jupiter's Rings

In 1979, the spacecraft Voyager 1 discovered that Jupiter has three thin rings circling it. The rings, which are too dark and dim to be seen by telescopes on Earth, are made of dust, probably thrown out by asteroid impacts on some of the small moons near Jupiter.

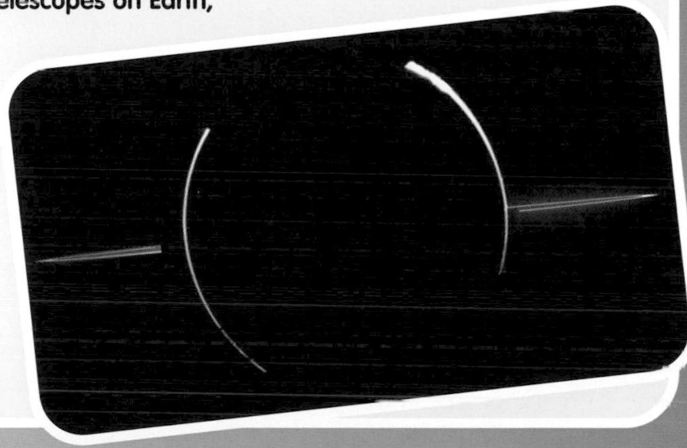

Four are large moons, discovered by Galileo when he first set his tiny telescope on Jupiter in 1610. These moons orbit the planet like a mini Solar System.

Jupiter's nearest large moon is Io (pronounced eye-oh), which is covered in active, erupting volcanoes. Hot gas and lava explode out to make hundreds of volcanoes. Io is the most volcanic place in the Solar System.

Next comes Europa, a moon made of rock but encased in ice. Cracks in the ice show there could be an ocean of liquid water underneath – and in that water there could be life. This makes Europa the most exciting moon in the Solar System. One day, a robot spacecraft may land on Europa, drill through the ice and launch a mini-submarine into the ocean to search for life.

The third moon, Ganymede, is the biggest moon in the Solar System. Larger than planet Mercury, Ganymede is also made of rock and covered in ice.

Finally comes Callisto, which is encased in ice and marked by thousands of craters. Perhaps the water on Callisto froze before the other moons and became scarred by meteorite impacts.

Saturn

Lord of the Rings

Saturn, the second biggest planet in the Solar System, is also the most beautiful because of its wonderful rings.

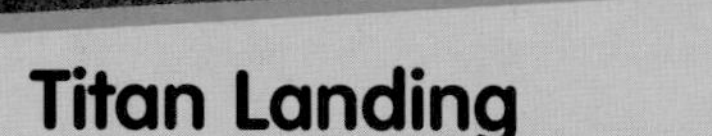

Titan Landing

In 2004 a spacecraft called Cassini began to orbit Saturn and its moons. In 2005 Cassini launched a small space probe called Huygens to land on Titan. The robot showed that Titan's air is yellow and the surface is covered in chunks of ice. Meanwhile, from orbit the Cassini spacecraft showed sand dunes on Titan's freezing deserts. Later pictures showed lakes of liquid methane.

The wide rings would stretch from Earth almost to the Moon. But they are thin, less than 0.6 miles/one kilometer deep. If Saturn's rings were as big as a football field, they would be as thin as a page in this book.

The rings are not solid. They are made of millions of pieces of ice and rock, in orbit around Saturn. Some pieces are as small as dust, other chunks are as big as a bus and others the size of buildings.

Some scientists believe Saturn did not always have rings. Perhaps, long ago, a moon made of ice and rock was in orbit around the planet, whose gravity ripped the moon apart and the pieces made Saturn's rings. Other scientists say the rings were made at the same time as the planet, so they have always been there.

Using telescopes on Earth, four different colored rings

Saturn Statistics

Size (diameter):	**75,000 miles/120,000 km**
Distance from sun:	**890 million miles/1,430 million km**
Day (rotation):	**10 hours, 14 minutes**
Year (time to orbit Sun):	**29 years**
Temperature:	**-290°F/-180°C**
Moons:	**57**

can be seen. Between some of the rings there are gaps, named after the astronomers who discovered them – the widest gap is called the Cassini Division. Spacecraft have taken close-up pictures, showing that they are made of thousands of tiny rings.

Saturn shows stripes, like Jupiter. But the stripes on Saturn are not as clear because it's covered in a high layer of fog. Sometimes bright white storms appear, but they are quite unusual. Winds on Saturn blow at 379 miles/600 kilometers per hour.

Saturn is not perfectly spherical as it is flatter at its poles and bulges out at the equator. Across the equator, Saturn measures 75,000 miles/120,000 kilometers, while from pole to pole it measures 67,000 miles/108,000 kilometers. This planet is squashed because it spins round quickly and is made of gas. The spin flings the gas out at the equator while the planet shrinks across its poles. All of the planets in our Solar System are squashed spheres, but Saturn shows it most.

Saturn has a large family of moons, and so far 57 have been discovered. Most are very small and are probably asteroids captured by the planet's gravity. Only seven moons are large and spherical in shape.

The largest moon, called Titan, is the second biggest moon in the Solar System and is the only moon with an atmosphere. Its air is mainly nitrogen, with some methane gas. Methane is found on Earth as natural gas and is used in gas fires and stoves. The air on Earth four billion years ago may have been like Titan's air.

Lightweight Planet

Saturn, which is made of gas, mainly hydrogen and helium, is the lightest, least dense planet in the Solar System. It would float on water – if you could find an ocean big enough!

Uranus, Neptune & Pluto

Uranus and Neptune are distant giant planets, similar in size but smaller than Jupiter and Saturn.

Uranus and Neptune both have an outer layer of hydrogen gas. Beneath the gas is a thick layer of ice and in the middle there is a small rocky core. The layer of ice gives the planets the nickname 'Ice Giants'.

Both planets are blue in color. This is the color of methane gas that freezes into a cloud above the hydrogen. Wisps of ammonia make white feathery clouds. Uranus and Neptune were both discovered using telescopes.

On March 13, 1781, astronomer William Herschel saw an object that did not appear on his star maps. It was a new planet, the first to be discovered using a telescope.

The planet was named Uranus, the name of the ancient god of the sky. Astronomers soon noticed that the new planet did not quite follow the orbit that they expected. Perhaps the gravity of another planet was affecting it. Uranus orbits the Sun on its side. Perhaps another planet collided with it long ago and knocked the planet over.

Both Uranus and Neptune have rings around them but they are too dim to be seen from Earth.

In 1846, a French scientist Urbain Le Verrier calculated where this unseen planet might be. An observatory in Berlin found the new planet on the first night of the search, close to where Le Verrier said it would be. It was named Neptune after the Roman god of the sea.

Uranus Statistics

Size (diameter):	**32,000 miles/51,000 km**
Distance from Sun:	**1,783 million miles /2,870 million km**
Day (rotation):	**17 hours, 14 minutes**
Year (time to orbit Sun):	**84 years**
Temperature:	**-353°F/-214°C**
Moons:	**27**

In 1930, the American astronomer Clyde Tombaugh discovered the ninth planet. It was named Pluto, a name suggested by 11-year-old Venetia Burney.

Pluto Statistics

Size (diameter):	**1,370 miles/2,200 km**
Distance from Sun:	**3,670 million miles/5,900 million km**
Day (rotation):	**6 days, 9 hours**
Year (time to orbit Sun):	**248 years**
Temperature:	**-380°F/-230°C**
Moons:	**3**

Pluto is so far away it's hard to see what it is like. It must be very cold, so most astronomers agree that it is made of rock and ice. We will know more in 2015, when the New Horizons space probe reaches Pluto.

In recent years, astronomers started to have doubts about Pluto. It was tiny, much smaller than our own Moon. Its orbit was strange, taking it nearly 50 times further from the Sun than Earth, then bringing it inside the orbit of Neptune. Was Pluto a 'proper' planet?

In 1978, a moon was found in orbit around Pluto and named Charon. With this moon, Pluto's status as a planet seemed more secure.

Then more objects like Pluto began to be found. Objects called Quaoar, Sedna and Eris were discovered. Eris (pronounced Ee-ris) was bigger than Pluto. Was Eris the tenth planet?

In 2006, astronomers from all over the world met together. After a lot of argument, they made a decision: Pluto was to be demoted. It was not to be called a planet any more. Instead, Pluto and Eris – and the largest asteroid, Ceres – would be called dwarf planets. So now, the Solar System contains eight planets and three dwarf planets.

Pluto and Eris belong to a part of the Solar System known as the Kuiper Belt, which contains lots of icy objects, including comets. And there are probably lots more dwarf planets like Pluto in this distant part of the Solar System.

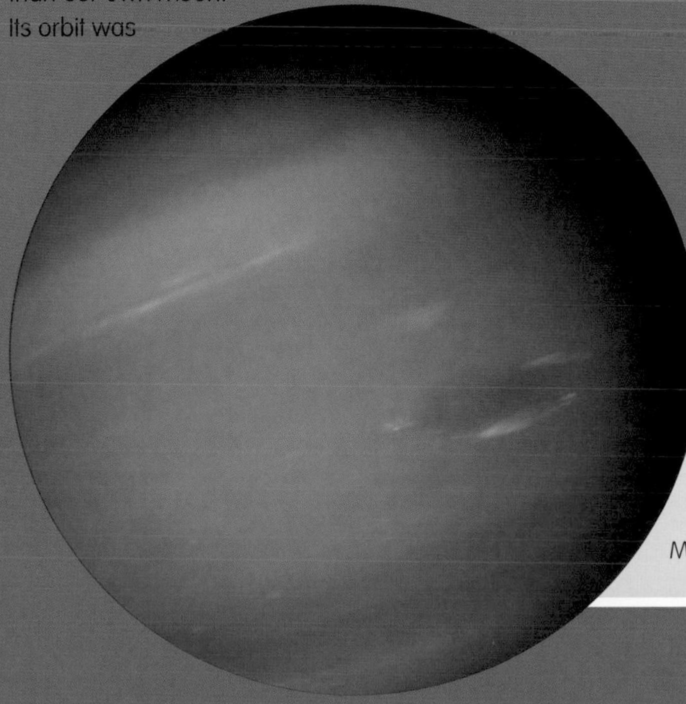

Neptune Statistics

Size (diameter):	**31,400 miles/50,500 km**
Distance from Sun:	**2,800 million miles/4,500 million km**
Day (rotation):	**16 hours, 7 minutes**
Year (time to orbit Sun):	**165 years**
Temperature:	**-364°F/-220°C**
Moons:	**13**

Asteroids

Leftovers of the Solar System

Between Mars and Jupiter, millions of rocks orbit the Sun. These are asteroids; pieces of rock that never came together to make a planet. The area in which they orbit is called the asteroid belt.

Asteroids can be as big as a house or the size of a mountain. Only three – Ceres, Pallas and Vesta – are over 300 miles/500 killometers wide. Ceres was promoted recently and is now known as a dwarf planet.

Car Crash

In 1992, a meteorite crashed into the Peekskill area of New York. It wrecked a car and the owner, Michelle Knapp, was understandably upset – at first. Then she sold the meteorite for $70,000/£35,000 and a TV company bought the car for $30,000/£15,000.

Asteroids were left over after the planets were made. The gravity of Jupiter stopped them from coming together to make a planet. If all the asteroids were squashed together, they would make a planet only as big as our Moon.

Sometimes, two asteroids crash into each other. One of them can be knocked towards the Sun and into a new orbit – and this new orbit might put the asteroid on a collision course with Earth.

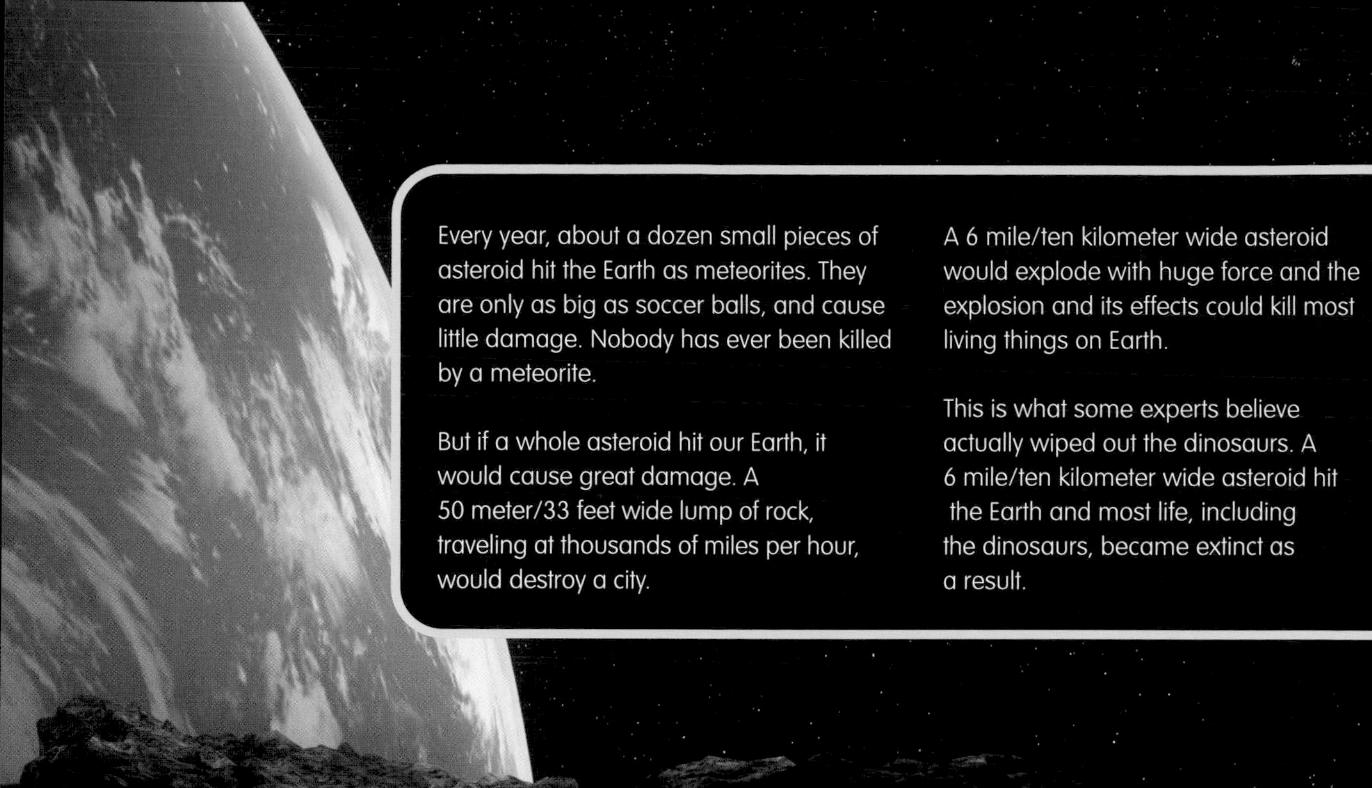

Every year, about a dozen small pieces of asteroid hit the Earth as meteorites. They are only as big as soccer balls, and cause little damage. Nobody has ever been killed by a meteorite.

But if a whole asteroid hit our Earth, it would cause great damage. A 50 meter/33 feet wide lump of rock, traveling at thousands of miles per hour, would destroy a city.

A 6 mile/ten kilometer wide asteroid would explode with huge force and the explosion and its effects could kill most living things on Earth.

This is what some experts believe actually wiped out the dinosaurs. A 6 mile/ten kilometer wide asteroid hit the Earth and most life, including the dinosaurs, became extinct as a result.

Comets

In the darkness beyond Pluto, millions of icebergs slowly orbit the Sun. These are the comets, dirty snowballs at the edge of the Solar System.

Halley's Comet

Halley's Comet is the most famous of all. It returns to our part of the Solar System every 76 years. Halley's Comet will be back in 2061.

Comets are mountains of ice with chunks of rock frozen inside. Outside, the ice is covered in dark dust. The area where comets orbit the Sun, called the Oort Cloud, is the edge of the Solar System, over 50,000 times further from the Sun than our Earth.

Sometimes two comets collide and one of them may be knocked in towards the Sun. As the comet comes near the Sun, the ice evaporates, making a huge ball of glowing gas. The Sun blows the gas away to make a blue gas tail and dust from the comet is blown away in a separate tail. Comet tails, which always point away from the Sun, can become millions of miles long and make a wonderful sight.

Gravity pulls the comet around the Sun then flings it into deep space, and some comets are thrown out of the Solar System and will never return. Others settle into a new orbit that brings the comet back many years later. Halley's Comet returns to the Sun every 76 years.

Four billion years ago, when the Earth was young, it was hit by many comets. The comet ice melted to make the seas and rivers we see now.

Some astronomers think comets did not just bring water. They believe that chemicals in comets were the starting materials for life.

Comet Crash

In 1994, a comet crashed into Jupiter after breaking into 20 pieces. When each piece hit Jupiter, it exploded and made a hole in the clouds, each about the size of our Earth. Jupiter is made of gas, so there were no craters, but if the comet had hit Earth, it would have destroyed much of our planet.

Meteors

On a clear night, you may be lucky enough to see a meteor. It looks like a falling star, flashing fast across the sky in a second or two.

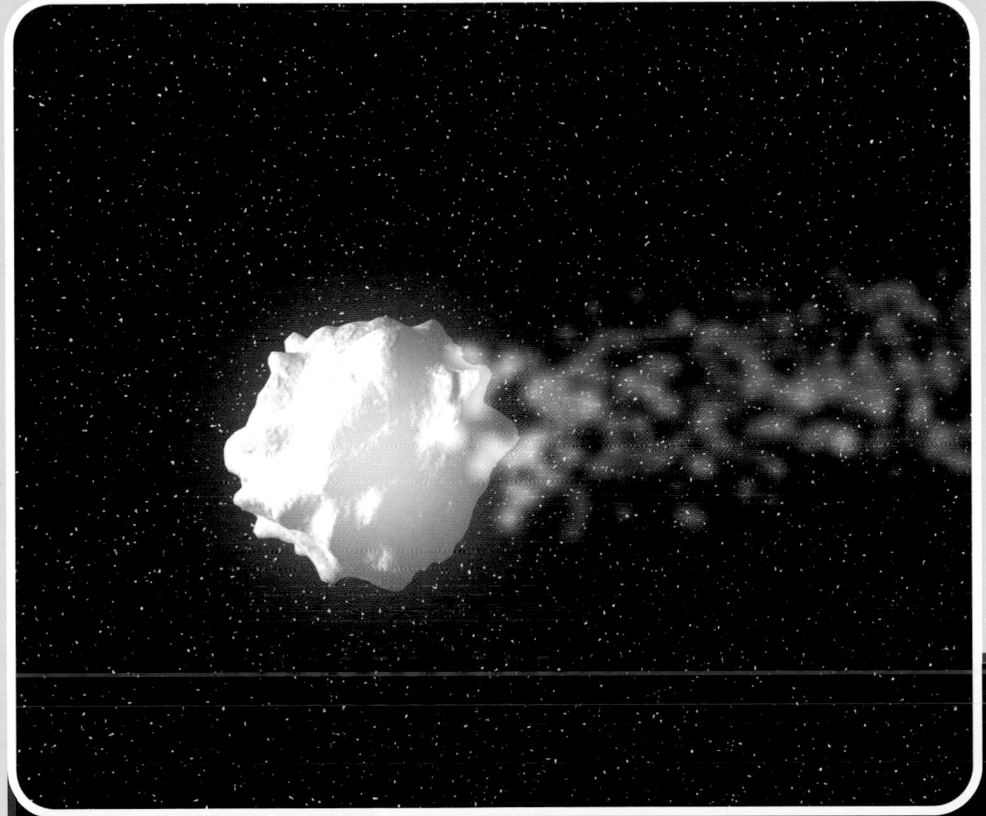

Lucky Meteors!

Long ago, people did not know what meteors really were, so they made up stories to explain them. Some said shooting stars were fights between angels and devils. Other people said they were the souls of dead people going to heaven. Even now, people think shooting stars are lucky. If you make a wish as you see a shooting star, it will come true.

Stars do not actually fall out of the sky. In fact shooting stars, or meteors, have nothing at all to do with stars. A meteor is a piece of space dust burning up in the air about 60 miles/ 100 kilometers high.

The dust, no bigger than a grain of sand, hits the air traveling faster than a bullet. At this speed, the grain of dust squashes the air in front of it. The squashed air becomes very hot and begins to glow – and the glow looks like a falling star. The heat from the hot air evaporates the dust as it falls through the sky.

A few shooting stars can be seen on any clear night. Then there are special nights when you can see lots of meteors. On these nights, our Earth crashes through a stream of space dust left behind by a comet. The space dust burns up to give lots of shooting stars – what we call a meteor shower.

There are several meteor showers, each happening on the same date each year. The best showers are around August 12 and December 13 . On these nights, if the sky is clear, you may see a shooting star every minute or so.

If you are very lucky you might see a meteor so bright it lights up the ground. This is a fireball, made by a piece of space dirt as big as a pebble.

Aurorae

Northern & Southern Lights

'I thought a house was on fire in the distance. The sky was glowing red. Then a green glow appeared. It moved among the stars like a curtain made of light. The colored lights glowed and danced for almost an hour.'

Light Height

The glow of an aurora comes from about 62 miles/100 kilometers high. At this height, the material from the Sun hits the air and makes the gas glow.

The lucky person who reported this event to a newspaper had not seen a fire. She had seen an aurora, the Northern Lights.

The aurora began on the Sun two days earlier. An explosion on the Sun blasted material into space – and toward the Earth. Our Earth's magnetism forced it toward the North and South Poles. When the material hit the Earth's air, it made the sky glow like a fluorescent light. Different gases in the air glowed different colors, red and green.

You have to be lucky to see an aurora, but two things can increase your chances. Aurorae are best seen near the North Pole or South Pole. So the further north or south you live in the world, the more likely you are to see an aurora. If you live in the Arctic Circle or in Antarctica, you will definitely see the lights.

You might still see an aurora where you live. When the Sun is very active, more explosions happen and there are more aurorae on Earth. Some may be big and bright enough to be seen in your town. The Sun will be very active in 2010 and 2011, so that's the time to look for these rare lights in the sky.

A bright aurora is a wonderful sky event, with curtains of colored light hanging in the sky, moving slowly and folding like a candle flame in a breeze. Sometimes, an aurora has streamers pointing upwards. Sometimes the light flickers, dims and brightens.

The main colors are green and blue: the green light comes from glowing oxygen and the blue comes from nitrogen. Sometimes, a rare red aurora is seen. The light comes from about 62 miles/100 kilometers above the ground, much higher than any clouds.

Many old stories were told about aurorae. In Scotland, the Northern Lights were said to be ghost dancers in the sky, while, in Greenland, they were messages from dead people. To the Vikings, they were lights from the shields of gods.

But we now know that material blasted out from the Sun makes the sky glow. It's not such a good story, but an aurora is still a brilliant sight to see.

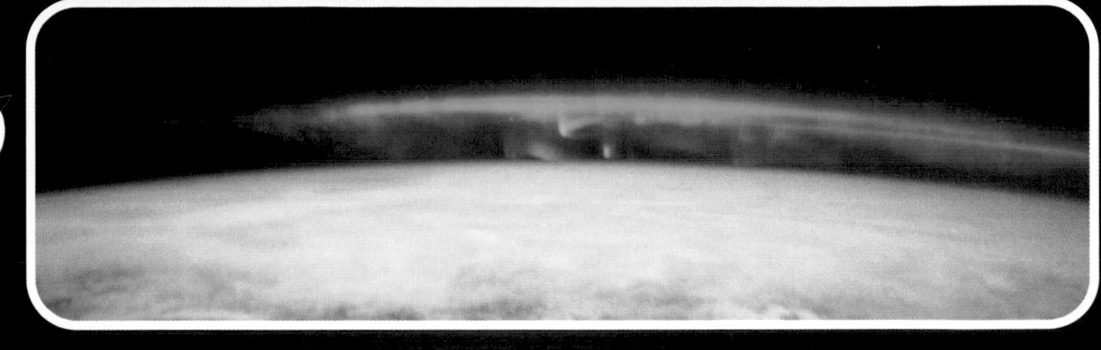

Auroral Rings

Aurorae are brightest in rings around the Earth's Poles. The brightest Northern Lights are seen at a latitude on Earth of 67 degrees north. Auroral rings have been photographed on Jupiter and Saturn.

Stars
The Jewels in the Sky

On a clear, dark night stars twinkle like jewels in the sky, they welcome us and invite us into the universe.

Your Star

From Earth, Sirius appears the brightest star in the night sky, because it's near us. It is eight light years away, which means its light takes eight years to reach us. When you see Sirius (known as the Dog Star), the light that goes into your eyes set off eight years ago.

Stars are so far away that measuring their distance in kilometers or miles gives numbers too big to understand. Instead, we use light years. A light year is the distance travelled by light in one year. Nothing can travel faster than light: 5,903,000,000,000 miles/9,500,000,000,000 kilometers in a single year. So one light year is 5.9 million million miles/9.5 million million kilometers.

A light year is a big distance. Even so, the stars are many light years away. The nearest, Proxima Centauri, is 4.2 light years away – that's almost 25 million million miles/40 million million kilometers! Other stars are tens, hundreds, even thousands of light years away.

When we look at the stars, we are looking a long way into space. We are also looking back in time. The light from stars takes many years to reach us, so we see the stars as they were many years ago. Some people say that the stars we see may not even be there any more. But don't worry, stars last for billions of years and they will be there long after we are gone.

Long ago, people made maps of the night sky, but they were no ordinary maps. They joined the stars together to make patterns, like playing dot-to-dot in the sky. The pictures they made were of people, animals and objects. The patterns, which we still use, are called constellations, of which there are 88.

The stars in a constellation are not really close together; they only make a pattern from our viewpoint on Earth. If you could travel to another star, the star pictures would not work.

As you look at the stars, you soon realize that they differ in brightness. Some appear dim while others look bright, for two different reasons. Some stars are truly bright, much brighter than our Sun, while others look bright only because they are closer to us.

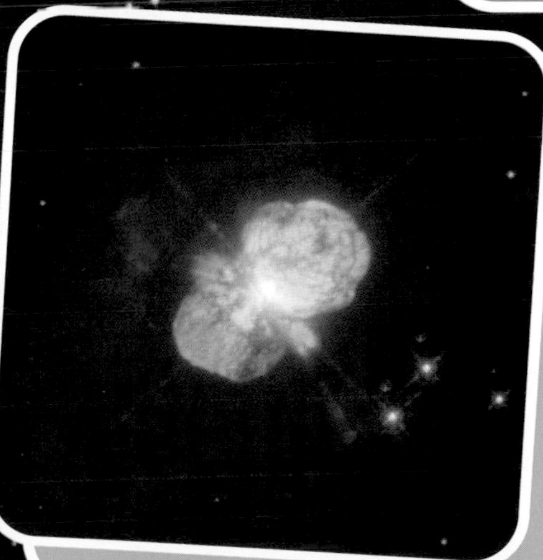

The Brightest Star

The truly brightest star is probably Eta Carina, which can be seen south of the equator. It is a superstar, 100 times bigger than the Sun and five million times brighter. There are probably less than 100 superstars like Eta Carina in our whole galaxy.

Star Looks

You might notice that different stars are different colors, and with binoculars the colors become stronger. There are four star colors – blue, white, yellow and red – and the color tells us how hot a star is. Blue stars are hottest, over 36,000°F/20,000°C. White stars are hot, about 11,000°F/10,000°C. Yellow stars like the Sun are medium temperature, around 5,400°F/6,000°C, while red (or orange) stars are least hot at 5,450°F/3,000°C.

A Star's Life Cycle

How stars live & die

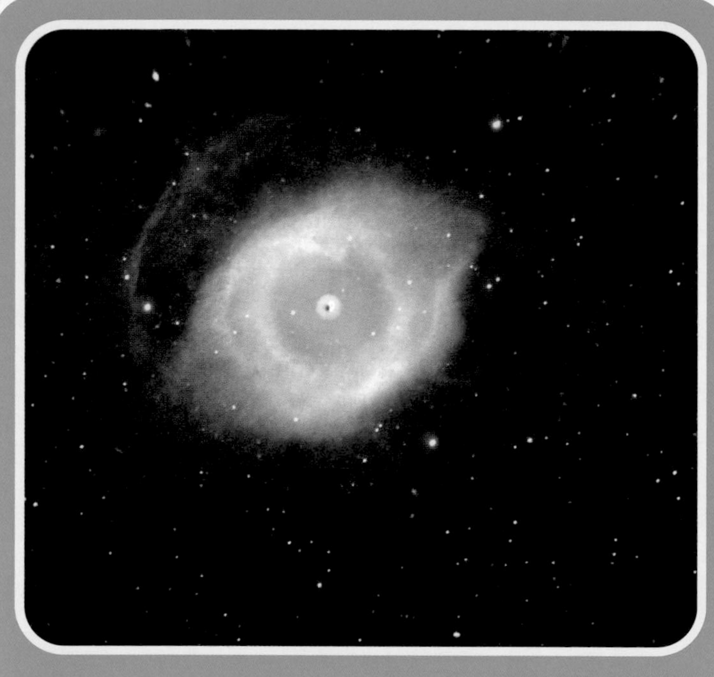

Nebula
Stars are born in nebulae, huge clouds of hydrogen gas. Gravity slowly pulls the gas together and part of the cloud collapses. The squashed gas becomes thicker and hotter. After a million years the gas in the middle is very thick and hot. When the temperature reaches 18 million degrees Fahrenheit/ 10 million degrees Celsius, the hydrogen explodes and a star is born. There are two sorts of stars that can be created during this process – either a sun-type star or a massive blue star. Follow the arrows to see how they evolve.

A star has two cycle options

Sun-type Star

Sun-type Star
Most stars are like the Sun. In the middle, hydrogen explodes and the energy makes the star shine. Only the middle reaches 10 million degrees. The gas around on the outside does not reach 10 million degrees Celsius, so it cannot explode. The gas in

Massive Hot Blue Star

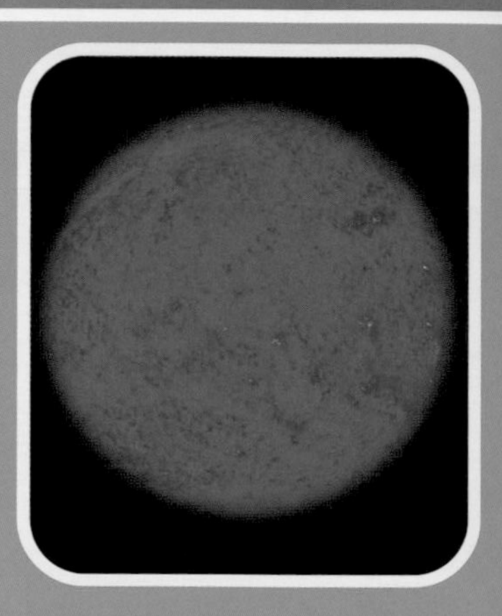

Massive Hot Blue Star
Some stars are made from big clouds of gas and contain lots of hydrogen. They become big, blue superstars. The core of a superstar is very hot and it burns up its supply of hydrogen quickly. Blue superstars live for a few hundred million years.

the shell is still collapsing under its own gravity. The explosion pushing out balances the force of gravity squeezing in, so the star settles to its size. The star has enough hydrogen to last for billions of years. Our Sun is has enough gas to last another five billion years.

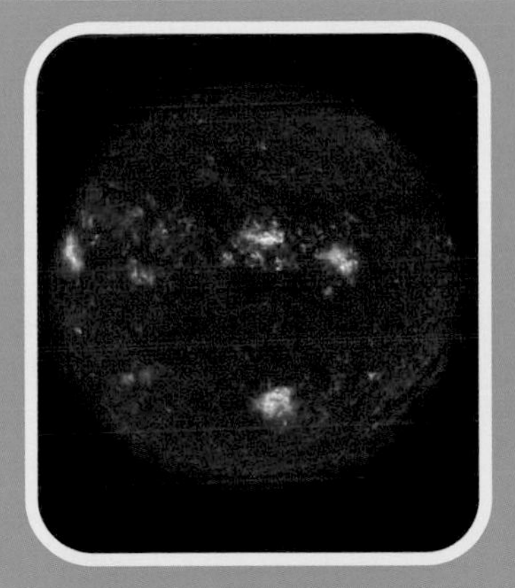

Red Giant
After billions of years, the star has burned up its hydrogen and swells up to 30 times its normal size. The surface cools and turns red: the star is dying as a red giant. In five billion years, our Sun will become a red giant and destroy the Earth.

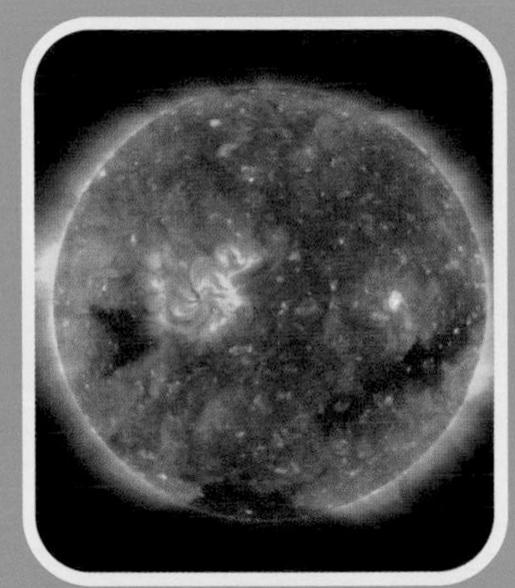

Outer Layers of Gas
The red giant becomes so big that it blows away its outer layers of gas. A ring of fire blows away from the dying star, burning up any planets. All that is left is the hot core. The ring of fire, which in a telescope looks round like a planet, is a planetary nebula.

White Dwarf
The hot core of the star is all that is left. It is a white dwarf, a dead star. It is squashed and very heavy – one teaspoonful of white dwarf would weigh a ton. The white dwarf will slowly cool and become a cold black ember. This is how most stars die.

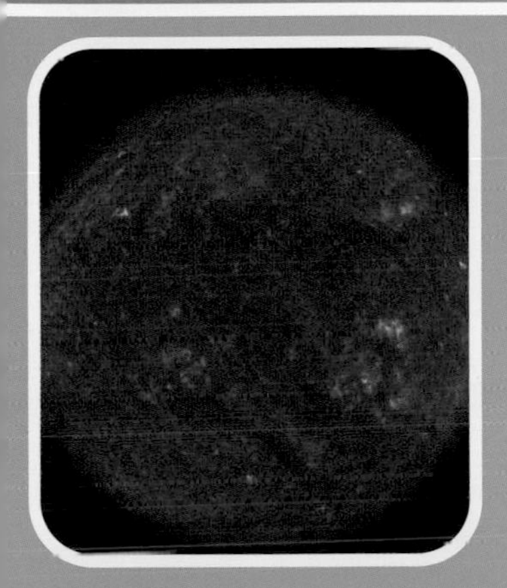

Red Supergiant
When the hydrogen in the core is used up, massive stars grow bigger. The heat is spread over a large surface and they go cool and red, becoming red supergiant stars. A supergiant is 100 times heavier than the Sun and 1,000 times bigger. The middle of the supergiant cannot hold up its massive weight for long. It is a starry time bomb.

Supernova
The gravity of the supergiant makes the star collapse suddenly. The middle is crushed into a tiny space and the outside layers fall inwards, then bounce off the squashed core. The star explodes in a supernova. A supernova explosion is as bright as 100 billion ordinary stars. Most of the star is blasted into space in a cloud.

Neutron Star
The middle of the exploded star is crushed by gravity. Three times the weight of the Sun is squashed into a ball only 12 miles/20 kilometers across. This is a neutron star, and it's very heavy – a teaspoonful of neutron star would weigh 100 billion tons! Some neutron stars cause bright flashes of light and are called pulsars.

Black Hole
The crushed core of the exploded star may weigh five times more than the Sun. Then gravity plays its final trick. The old core is squashed into a space smaller than this full stop. This tiny object has gravity so great that nothing can escape from it, not even light. A ball of darkness is made. This is what is known as a black hole.

Nebulae
Clouds in Space

The space between stars is not quite empty. Telescopes show fuzzy glows in the dark, night sky. These are nebulae, clouds of gas and dust between the stars. Nebulae, which are some of the most colorful objects in space, come in different kinds.

Bright Nebulae
Bright nebulae are made of hydrogen gas mixed with a little helium. Stars inside the cloud give the energy to the gas, which glows like a fluorescent light. Hydrogen gas glows red, so photographs of nebulae appear a reddish color. Some bright nebulae are huge, spreading out for many light years between the stars. Inside bright nebulae, gravity is at work. It pulls the hydrogen gas together and the gas collapses on itself, eventually making new stars. Dust in a nebula can make new planets.

Orion Nebula
The most famous bright gas cloud is the Orion Nebula, in the constellation of Orion the Hunter. It is the brightest part of a gas cloud that covers the whole of Orion. The nebula glows brightly because four hot stars light it up. The stars make a shape called a trapezium, like a squashed square.
Inside the Orion Nebula, new stars are being born. Some of these baby stars have a ring of dust around them, and this dust may be making planets.

Horse-head Nebula
The best-known dark nebula is in the constellation Orion. Thick, cold dust blocks out the light of a bright nebula, and this dust has swirled around to make the shape of a horse's head. Energy from nearby stars is blowing the dust away and changing the horse-head shape.

Dark Nebulae
Sometimes a dust cloud is so thick that it blocks out the light of stars behind it. The dust looks like a dark hole in the sky and it is called a dark nebula.

Planetary Nebulae
When old stars begin to die, they blow away their outer layers of gas. This gas makes a ring of fire that looks like a circular glow. Their round shape makes them look a bit like planets in a telescope, so in days gone by they were called planetary nebulae. In fact they have nothing to do with planets – they are dying stars.

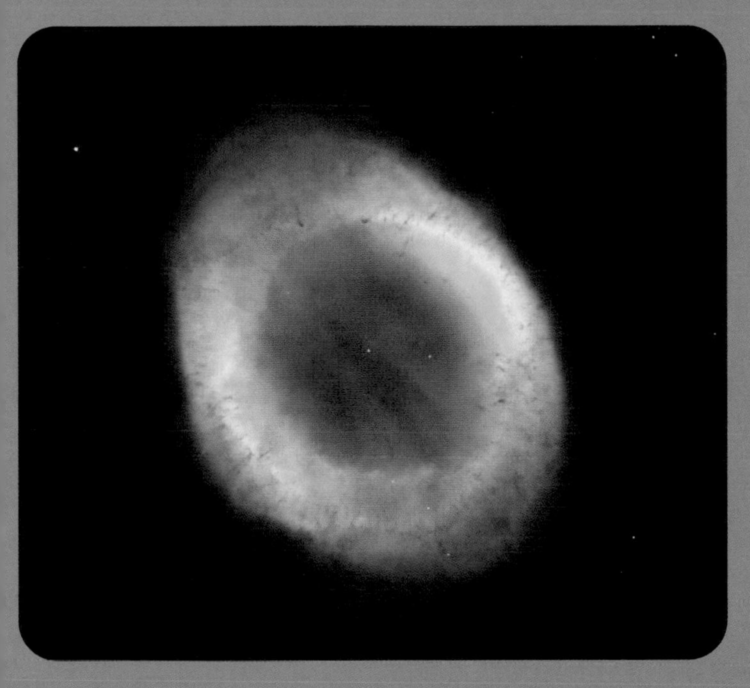

Pillars of Creation

One of the best pictures from the Hubble Space Telescope showed piles of dust in a nebula. They were called 'The Pillars of Creation' because stars and planets were being made inside. Gas comes together to make stars, and dust sticks together to make planets.

Supernova Remnant
A huge superstar dies in a massive explosion called a supernova. The explosion blasts gas and dust into space and this makes a nebula called a supernova remnant. The brightest is the Crab Nebula in the constellation of Taurus the Bull. This is the remains of a star that was seen to explode in the year 1054.

Reflection Nebulae
Some clouds contain a lot of dust, and starlight bounces off the grains of dust like smoke in a beam of light. The reflected starlight glows blue to make a reflection nebula. These nebulae are much fainter than glowing bright nebulae because the light is only reflected.

M for Messier

Charles Messier was a French astronomer who lived over 200 years ago. He used his telescope to find comets, which look like fuzzy glows in the sky. But nebulae also look like fuzzy glows. Messier became fed up with believing that he had found a new comet, only to find it was a nebula, so he made a list of nebulae and where they were in the sky.
We now use Messier's list for nebulae, star clusters and galaxies. There are 110 objects in his list, from M1 to M110. The Orion Nebula is M42.

Galaxies
Cities of Stars

Stars collect together in huge star cities called galaxies. They contain billions of stars and clouds of gas and dust, with gravity holding the stars, gas and dust together. Galaxies are millions of light years apart, with only empty space between them.

Hubble Deep Field

The Hubble Space Telescope has taken a picture of the furthest galaxies ever seen. The most distant galaxies in the picture are 12 billion light years away. Light from these galaxies set off before the Earth was made! It took Hubble ten days to collect the light from these star cities.

A Universe of Galaxies

There are around 150 billion galaxies in the universe. A galaxy holds about 100 billion stars. Now you can work out how many stars there are in the universe – it is about 150,000,000,000 x 100,000,000,000.

Galaxies come in different shapes and sizes, and the most spectacular are spiral galaxies. These have a bright, round middle, glowing yellow with old stars. Spiral arms sprout from the middle with hot blue stars and red nebulae. The arms are wide and flat, like an enormous dinner plate in space. The nearest spiral galaxy is the vast Andromeda Galaxy, which holds one thousand billion stars.

Some spiral galaxies have a long bar of old stars in the center, with just a couple of spiral arms attached. These are called barred spiral galaxies.

Other galaxies are quite different, being round or oval in shape and having no spiral arms at all. They glow yellow and contain mostly old stars. Another name for an oval is an ellipse, so these are called elliptical galaxies. Most elliptical galaxies are small, contain only a few million stars, and are called dwarf elliptical galaxies. A few, called giant ellipticals, are huge, much bigger than the biggest spirals.

The third type of galaxy has no particular shape, and these are known as irregular galaxies.

Our own galaxy is called the Milky Way. Although we live inside the Milky Way, we cannot see all of it. We can see the thin spiral arms making a band of stars across the sky. Ancient people thought it looked like milk spilt across the sky, and that is how our star city was named. Radio telescopes can show us the shape of the Milky Way, and it is a spiral, like Andromeda. But it is much smaller, with about 200 billion stars.

Our galaxy is about 100,000 light years across, and our Solar System is about two-thirds the way from the middle of the Milky Way. At the center of our galaxy there seems to be a huge black hole, which was made from exploded superstars. In fact, it seems that there is a big black hole at the middle of every galaxy.

Just like stars, galaxies are not spread evenly through space. They collect together in clusters, and clusters of galaxies are collected in superclusters! Galaxy superclusters are the biggest things in the universe.

Colliding Galaxies

In some galaxy clusters, galaxies can be seen colliding together, but the stars are so far apart that they rarely hit each other. The galaxies pass through each other like ghosts in the night.
In our cluster, the Milky Way and Andromeda are moving towards each other at 500,000 km per hour/310,000 miles per hour. In five billion years, the two galaxies will collide. Now that would be a sight to see.

People in Space

Small Steps & Giant Leaps

On April 12, 1961, an unknown Russian cosmonaut named Yuri Gagarin blasted off in the Vostok 1 spacecraft to orbit the Earth. He returned 108 minutes later as the most famous person on the planet. Yuri Gagarin had become the first person in space.

Just eight years later, on July 21, 1969, Neil Armstrong and Buzz Aldrin stepped onto the Moon and into history. They had become the first people to visit another world.

Between 1969 and 1972, 12 American astronauts visited the Moon. It seemed that from these first small steps the next giant leap to Mars would soon be made, but it didn't happen.

People on Earth began to question the cost and value of sending astronauts to other planets. The big, exciting missions into space were at an end.

Return to the Moon

By 2020, humans should be back on the Moon. NASA's Constellation project may build new rockets called Ares and a new space module called Orion, which will be able to carry four astronauts. In 2014, Orion will be tested by a visit to the space station and, in 2020, an Ares rocket will carry its astronauts to the Moon.

In the 2020s, astronauts will build a base on the Moon and, if all goes well, an Ares rocket will take Orion and its astronauts to Mars in the 2030s. Humans will land on Mars during your lifetime.

Weightlessness

Astronauts float around in the space station because they are 'weightless', but it's not lack of gravity that makes them weightless. There is gravity in space, and in fact gravity holds the space station in orbit.

Astronauts are in fact falling towards the Earth – but never getting anywhere. This free fall makes them feel weightless. Astronauts call this microgravity.

Space Tourist

You may also have the chance to travel into space. Already, space tourists have visited the International Space Station. The cost is about $20 million/£10 million for a ten-day stay in space! Now some companies are looking at cheaper flights. For a few hundred thousand dollars, you may be able to take a short hop in a rocket and experience a few weightless minutes, just like an astronaut.

Now, in a new century, there are plans for the big missions to return. After 1972, the Russians and Americans built space stations to orbit the Earth, giving cosmonauts and astronauts the chance to spend a long time in space. The Americans built the Space Shuttle, a spacecraft that took off like a rocket and returned to Earth like an airplane. This was the first spacecraft that could be used over and over again. Now the Space Shuttle and Russian Soyuz rockets visit the International Space Station, ISS, in orbit 225 miles/360 kilometers above the Earth.

ISS, which has been built in orbit, is home to three astronauts at a time and each crew spends several weeks working in space.

The International Space Station is the biggest human construction in space. It measures 240 feet/73 meters across its solar panels, is 144 feet/44 meters long and 92 feet/28 meters high.

It takes 91 minutes to orbit Earth, traveling at a speed of almost 17,400 miles per hour /28,000 kilometers per hour.

But there are hidden dangers for people who spend a long time in space. Astronauts' muscles and bones become weak in their weightless environment and, outside our atmosphere, radiation from explosions on the Sun might hurt people.

We need to know about these dangers because the next steps into space may be taken soon.

Robots in Space

Explorers of the Universe

On October 4, 1957, the world listened in wonder to a radio signal from space. It came from Russia's Sputnik 1, the first spacecraft to orbit the Earth.

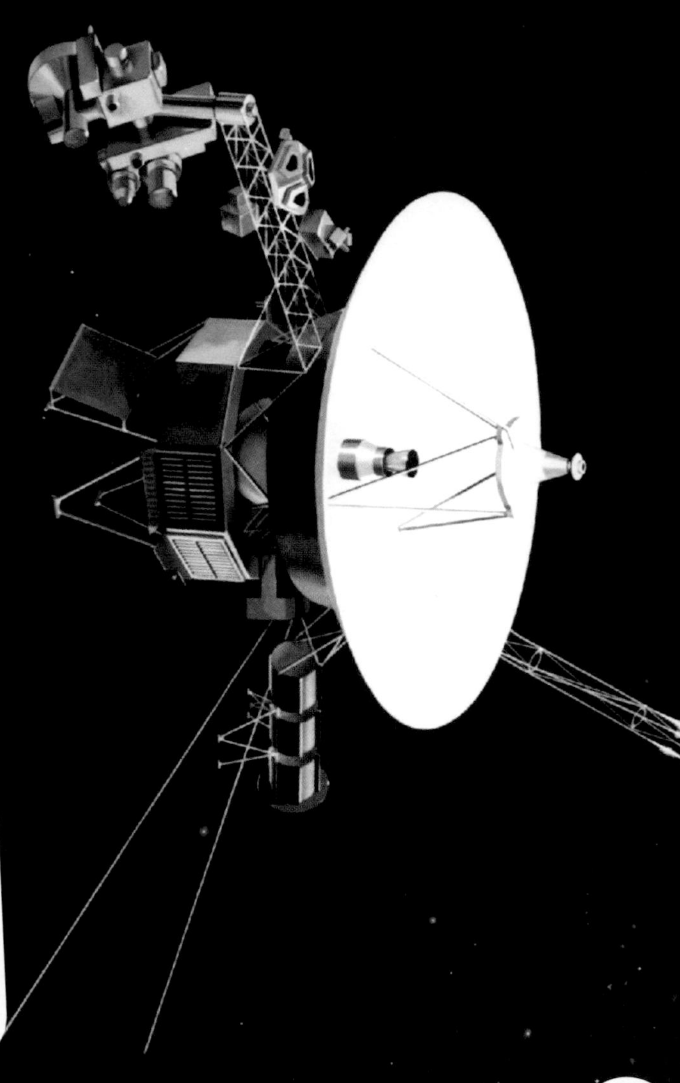

Pluto and Beyond

The only planet that has not been visited by robots is Pluto. But, even though Pluto is now a dwarf planet, there is a mission on its way there. New Horizons was launched in January 2006 and flew by Jupiter in February 2007. It will arrive at Pluto in July 2015. It was the fastest spacecraft launch ever, reaching a speed of over 22,000 miles per hour/36,000 kilometers per hour.

Voyagers Now

The Voyager spacecraft are leaving the Solar System on a journey to the stars. Voyager 2 is beyond the orbit of Pluto, while Voyager 1 is further away, about 100 times further from the Sun than Earth. In 40,000 years they will pass nearby stars. Both spacecraft carry gold discs with recordings of sights and sounds of Earth. If aliens find a Voyager, they will discover what we are like!

Sputnik 1 was a tiny satellite, about the size of a basketball. Since 1957, thousands of bigger satellites have been launched into orbit around the Earth. We now use satellites for TV, phone calls, navigation and weather forecasts. They also help us to find out about space.

The most famous astronomy satellite is the Hubble Space Telescope. Hubble orbits 370 miles/590 kilometers above the Earth. Outside our wobbly, dirty air, Hubble has a clear view into deep space. Its eight feet/2.4 meter-wide telescope has given us wonderful views of stars, nebulae and galaxies. It has changed our ideas about the universe.

Robots have traveled far into the Solar System, to places too far away and too dangerous for humans. Their pictures and data have changed our ideas about the planets.

The most spectacular robot mission was Voyager. In 1977, two robot spacecraft set off to explore the outer planets and their moons. Both spacecraft flew by Jupiter and Saturn. Voyager 1 visited Saturn's moon Titan, and was then flung into deep space. Voyager 2 continued its epic trip to Uranus and finally reached Neptune in 1989.

The Voyagers paved the way for the Galileo spacecraft to orbit Jupiter and its moons, then the Cassini mission to Saturn and its satellites. Every mission added new information. We learned more about the Solar System in 20 years than in the whole of human history that had gone before.

In the 1960s, robots landed on the Moon before people set foot there, and later robots have landed on Venus and Mars. The Russian Venera landers, the strongest spacecraft ever built, sent back pictures of the surface of Venus but survived only two hours in the hot, acid, crushing atmosphere. The Mars landers Viking, Pathfinder and now Spirit and Opportunity have explored the surface of Mars. Like the robots on the Moon, they have prepared the way for humans to follow in the future.

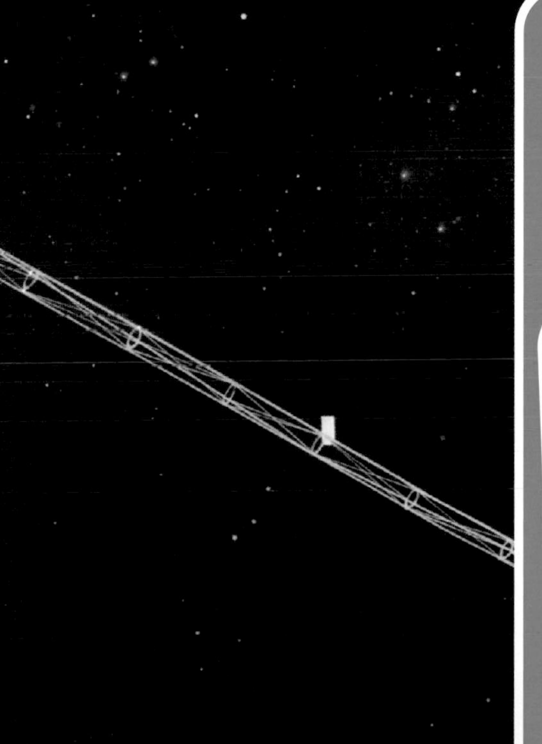

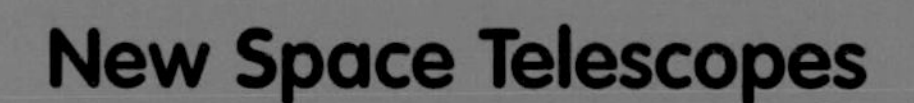

New Space Telescopes

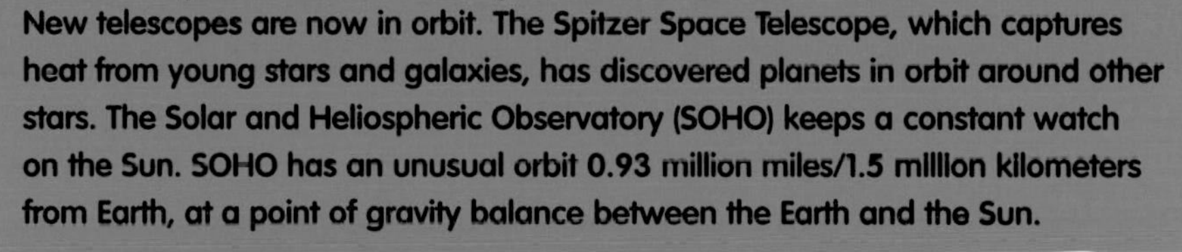

New telescopes are now in orbit. The Spitzer Space Telescope, which captures heat from young stars and galaxies, has discovered planets in orbit around other stars. The Solar and Heliospheric Observatory (SOHO) keeps a constant watch on the Sun. SOHO has an unusual orbit 0.93 million miles/1.5 million kilometers from Earth, at a point of gravity balance between the Earth and the Sun.

Northern Hemisphere

Winter Sky Window

December to February

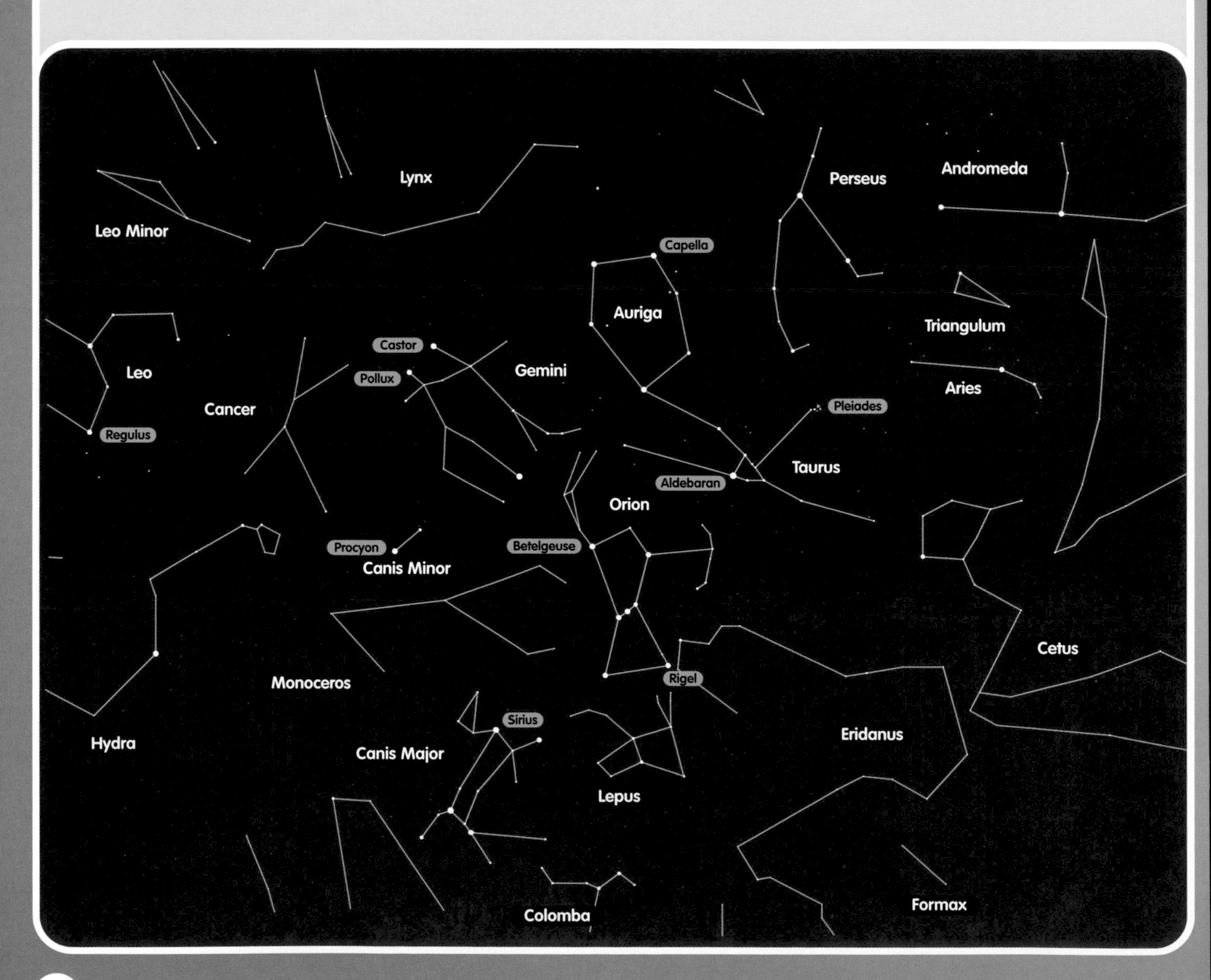

The maps for the Northern Hemisphere show the night sky as seen by people who live North of the equator. So if you live in the UK or in North America for example, use these charts.

Using The Sky Window

To find south, you can use a compass, or you can remember where the Sun has set, which is west. Keep the sunset on your right, look ahead, and you will be looking south.

Winter is a great time to look at stars. The nights are long, so go out early, but wrap up well. Thick socks, gloves and a woolly hat will keep you warm. Remember to always ask permission from an adult before venturing out.

Skylights

Orion the Hunter will show us the sky. First, find three stars in a straight line. This is Orion's belt. Above the belt, two bright stars make Orion's shoulders, and, below the belt, two more stars make his knees.

The top left star is Betelgeuse, a red supergiant. Betelgeuse is 800 times wider than the Sun and 40,000 times brighter. One day, it will explode in a supernova and leave a black hole in space. Luckily for us, Betelgeuse is a safe 420 light years away!

The star at bottom right is Rigel, a hot blue superstar. Rigel shines with the light of 60,000 Suns from a distance of 800 light years.

Follow the line of the belt down to the left. The bright twinkling star is Sirius, the dog star, the brightest star in our night sky. It's not really as bright as Betelgeuse or Rigel, but it's only eight light years away.

Look above Sirius and the next bright star is the puppy, Procyon. Go right of Procyon and you are back to Betelgeuse. Sirius, Procyon and Betelgeuse make a neat triangle: the Winter Triangle.

Above Procyon you will find the twin stars of Gemini. The upper twin is Castor and the lower star is Pollux.

Gemini is the place for excitement around December 13 each year, when there are lots of shooting stars coming from this part of the sky – the Geminid meteor shower.

Now go back to Orion's belt. Follow the line of the belt up to the right. The orange star you see is Aldebaran, eye of Taurus the Bull.

Keep going right until you see a fuzzy patch of stars. Now you are looking at a star cluster called the Pleiades, the Seven Sisters.

Finally, look overhead. The bright yellow star is Capella, belonging to the constellation Auriga the Chariot Driver. Capella's light comes from two stars so close together that our eyes see it as only one.

Observing Checklist

With your eyes only:

- Learn the shapes of Orion, Taurus, Gemini and Auriga
- Learn the stars Betelgeuse, Rigel, Sirius, Procyon, Aldebaran, Castor, Pollux and Capella
- Count how many stars you can see in the Seven Sisters Cluster

With binoculars:

- Look at the colors of these stars – Betelgeuse, Rigel, Sirius, Procyon, Aldebaran, Castor, Pollux and Capella

To find out more:

- Look on pages 74 and 75 for Constellation Close-Ups on Orion and Taurus

Northern Hemisphere

Spring Sky Window

March to May

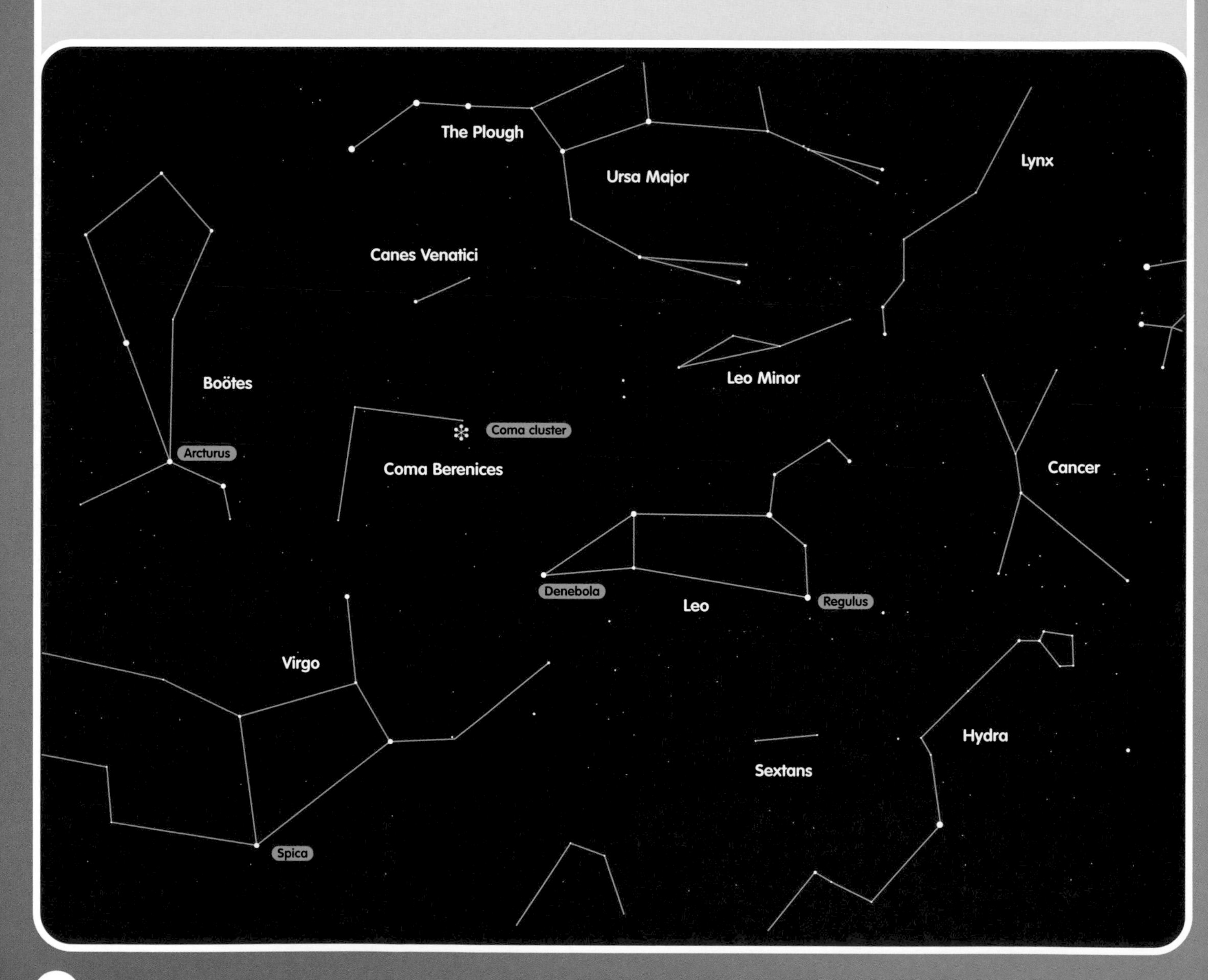

Observing Checklist

With your eyes only:

- Learn the shapes of Leo, Virgo, Boötes and the Plough
- Learn the stars Regulus, Arcturus and Spica
- Find the Coma star cluster

With binoculars:

- Look at the colors of these stars – Regulus, Arcturus and Spica
- You can see the Coma star cluster

To find out more:

- Look on pages 76 and 77 for Constellation Close-Ups on Leo and Boötes

How to Use the Sky Window

Remember where the Sun has set, roughly in the west. Keep the sunset on your right and look ahead and you will be looking south.

Skylights

The great constellation of spring is Leo the Lion, prowling the sky in the south.

To find Leo, look for stars making a big hook in the sky, like a question mark drawn backwards. The bright star at the bottom of the hook is Regulus. Its name means 'King Star' because it is near to the path of planets across the sky. Regulus is 150 times brighter than our Sun and 80 light years away. To the left of Regulus, other stars make the lion's body and his tail is a star called Denebola.

In 2008 and 2009, Leo will seem to have an extra 'star'. But the star will really be the planet Saturn, so it will be a great time to look at those wonderful rings.

Now look above Leo. Almost overhead is the Plough, a saucepan shape of seven stars. Some people think it's more like a spoon, so it is sometimes called the Big Dipper. Now is the best time to look at the Plough because it is so high in the sky.

When you see the saucepan shape, follow its handle down to the left, eastwards. The next bright star is Arcturus, the fourth brightest star in the sky. This is an old red giant, about 40 light years from Earth.

Above Arcturus, look for a kite-shape of faint stars. This is the shape of a man called Boötes (pronounced Boo-ay-tays) the Shepherd.

Between Leo and Boötes you might see the dim stars of Coma Berenices. When you try to look straight at it, the patch disappears! The stars make the Coma star cluster, and we see it better using the edges of our eyes.

Below Arcturus, quite low in the sky, the next bright star is Spica. This is a superstar, 14,000 times brighter than the Sun. Spica is the brightest star in Virgo and shines from a distance of 260 light years.

Look above Spica and you might see other stars making a dim letter 'V'. Virgo has her initial in the sky! Saturn will have moved in front of Virgo by 2010 and 2011.

When we look towards Virgo, we are looking out of our galaxy, the Milky Way. In this part of the sky, big telescopes have found hundreds of faint galaxies far away.

They belong to the Virgo cluster of galaxies, and our own Milky Way and galaxies around us belong to the Virgo supercluster of galaxies.

When we look towards Virgo, we are looking out of our own galaxy, the Milky Way.

Northern Hemisphere

Summer Sky Window

June to August

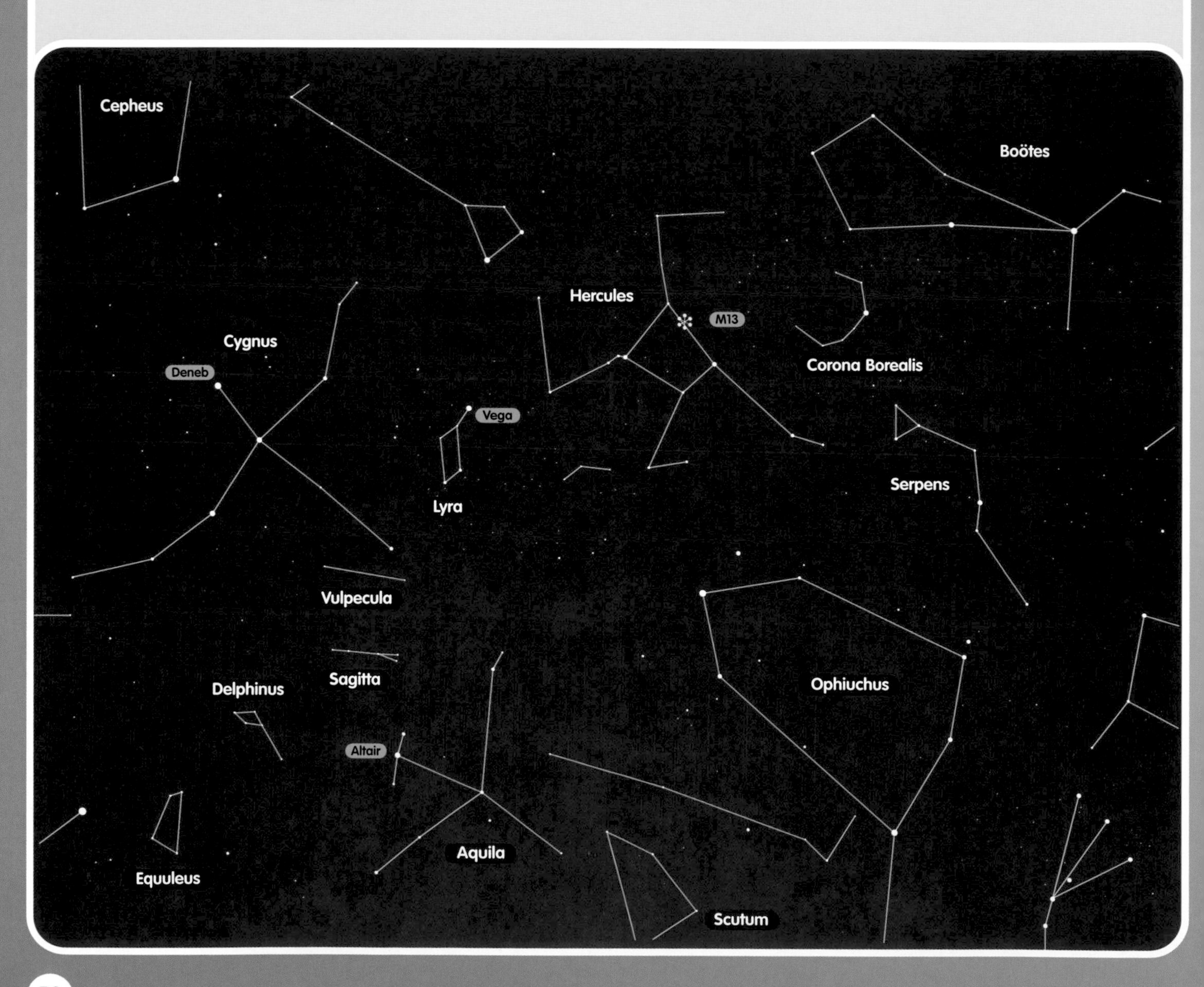

Observing Checklist

With your eyes only:

- Learn the shapes of Cygnus, the Summer Triangle and Hercules
- Learn the stars Vega, Deneb and Altair
- Find the tiny star sign of Delphinus the Dolphin
- Look for the cloudy band of the Milky Way

With binoculars:

- Observe the colors of these stars – Vega, Deneb and Altair
- Find the great Globular Cluster in Hercules

With a telescope:

- Observe the Globular Cluster M13

How to Use the Sky Window

To find south, remember the direction of sunset, which is roughly west. Keep the sunset on your right and look ahead. You are now looking south.

Skylights

Astronomers have to stay up late in summer. Long days and short nights mean that the sky goes dark late in the evening.

When it does go dark, three bright stars can be seen before the rest. They make a huge triangle called the Summer Triangle, and that's where to start our summer observations.

At top right of the triangle is Vega, which looks bright because it is fairly close to us – only 25 light years away.

Left of Vega is superstar Deneb, much brighter than Vega but much further away. It shines 70,000 times brighter than the Sun from 1,800 light years away.

The third star of the triangle, Altair, is below the other two. Altair is only ten times brighter than the Sun and is closer than Vega, only 16 light years away. Stars around Altair make the shape of Aquila the Eagle – and there is another big bird nearby.

If you go back to Deneb and look below it, into the triangle, you will see four stars in the shape of a cross. Its nickname is the Northern Cross, but its actual name is Cygnus the Swan. Deneb is the swan's tail and the other stars make its wings and long neck. The swan is flying down through the triangle.

There is a little constellation just to the left of Altair. Look for a tiny diamond of stars, with another star underneath. This is the shape of Delphinus the Dolphin, leaping out of the sea. To the right of Vega, you can see a dim squashed square of stars. This is the body of Hercules, the world's strongest man. Look between the two stars to the right of the squashed square. About one-third of the way down, you might see a misty star. In binoculars it's a fuzzy glow, but with a telescope it becomes a ball of stars.

This ball of stars is the Great Globular Cluster M13. About one million stars are packed closely together about 25,000 light years away.

During 2008 and 2009, the great planet Jupiter will be low in the south during summer. Every year around August 12, meteors flash across the sky. They belong to the Perseid meteor shower, one of the best displays of the year so, if it's clear on August 12, stay up to watch for the shooting stars.

Every year around August 12, meteors flash across the sky. They belong to the Perseid meteor shower, one of the best displays of the year.

Fall/Autumn Sky Window

September to November

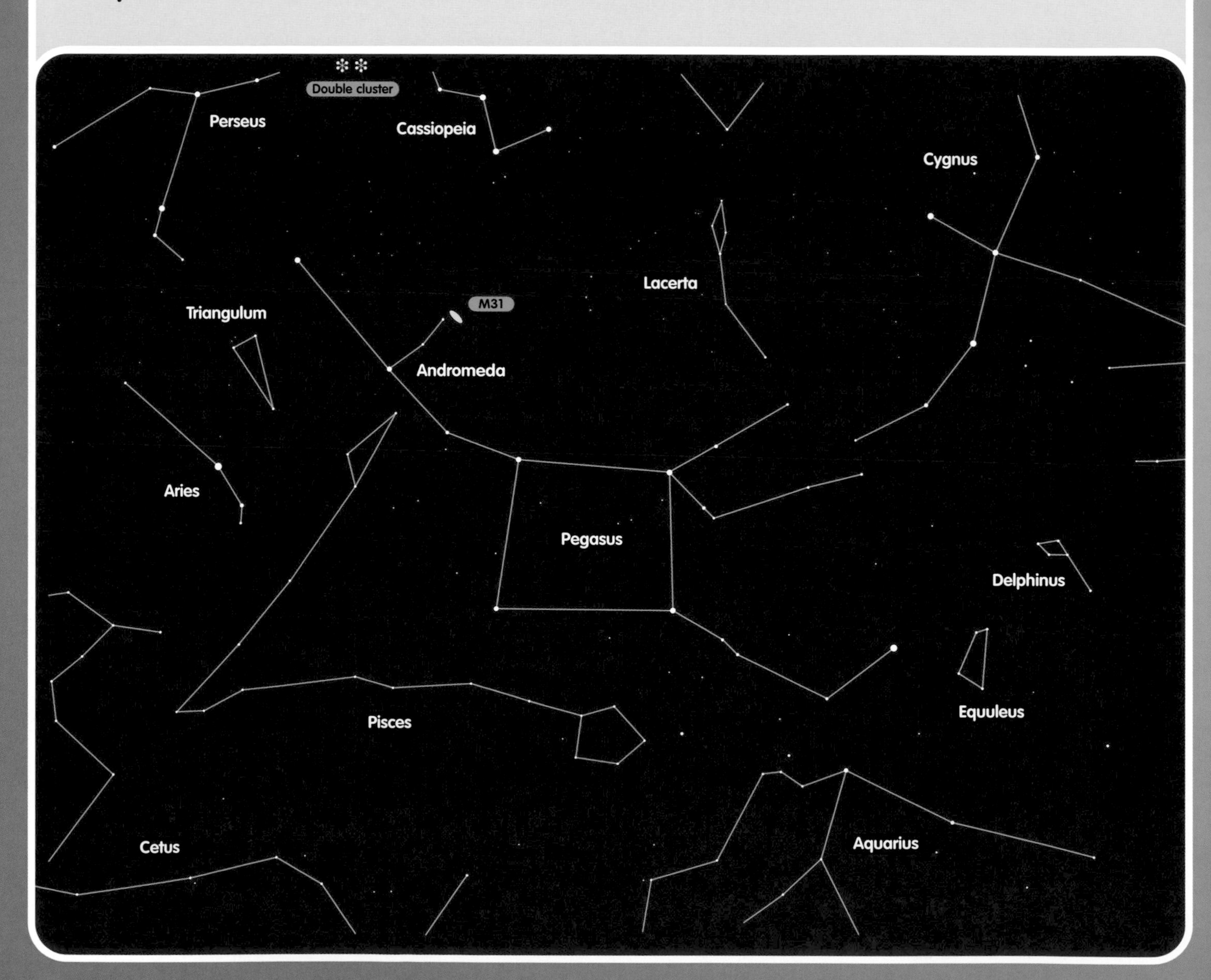

Observing Checklist

With your eyes only:

- Learn the shapes of Pegasus, Andromeda, Cassiopeia, Perseus and Cetus
- Find the Andromeda Galaxy, M31
- Find the Double Cluster

With binoculars:

- Look for the Double Cluster between Perseus and Cassiopeia

To find out more:

- Ov pages 78 and 79 for Constellation Close-Ups on Pegasus and Andromeda

How to Use the Sky Window

To find south you can use a compass, or you can remember where the Sun sets, in the west. Keep the sunset on your right and look ahead to the south.

Skylights

As fall/autumn arrives, the sky becomes darker earlier, but the nights are still warm. It's a good time to go outside and look at the stars.

Looking south, you will see a big square made of four stars. Although they are not really bright, the square stands out. Your eyes and brain are good at drawing shapes in the sky. The four stars make the Square of Pegasus. At bottom right, more stars make his neck and bend up for his head. The famous flying horse is actually flying upside down!

The fall/autumn sky will have an extra bright 'star' from 2010 to 2012: the planet Jupiter will cross the sky below the square, through the fishy stars of Pisces.

From the top left of the square, look for a line of four more stars. This is the slim body of Andromeda the Princess. Just above the line of Andromeda, your eyes may pick out a hazy glow – this is the furthest thing you can see with your eyes. The glow comes from the Andromeda galaxy, code number M31 and an incredible 2.5 million light years away. The light that comes into your eyes left the galaxy over two million years ago, before there were people on Earth!

In the old story, Andromeda was rescued from a sea monster, which still lurks in the sky with the name of Cetus. Look below, to the left of the square, and there he is. Dim stars make his body, then up to the left stretches his long neck and round head. You really need a good imagination to draw these pictures!

Overhead in the fall/autumn is Cassiopeia the Queen, mother of Andromeda. Her five brightest stars make a letter 'W' in the sky. In the story, Cassiopeia annoyed the god of the sea, who sent a sea monster to gobble up poor Cassiopeia.

Andromeda's stars point to the hero who rescued her. Follow her line of stars from the square, and the next group of stars is Perseus.

Between Cassiopeia and Perseus, you might notice a ghostly glow which, with binoculars, becomes two bright clumps of stars. This is the famous Double Cluster. You can explore each cluster – code named NGC 884 and NGC 869 and about 8,000 light years away – separately, with a telescope.

Andromeda's stars point to the hero who rescued her. Follow her line of stars from the square, and the next group of stars is Perseus.

Sky Window

All-year Stars

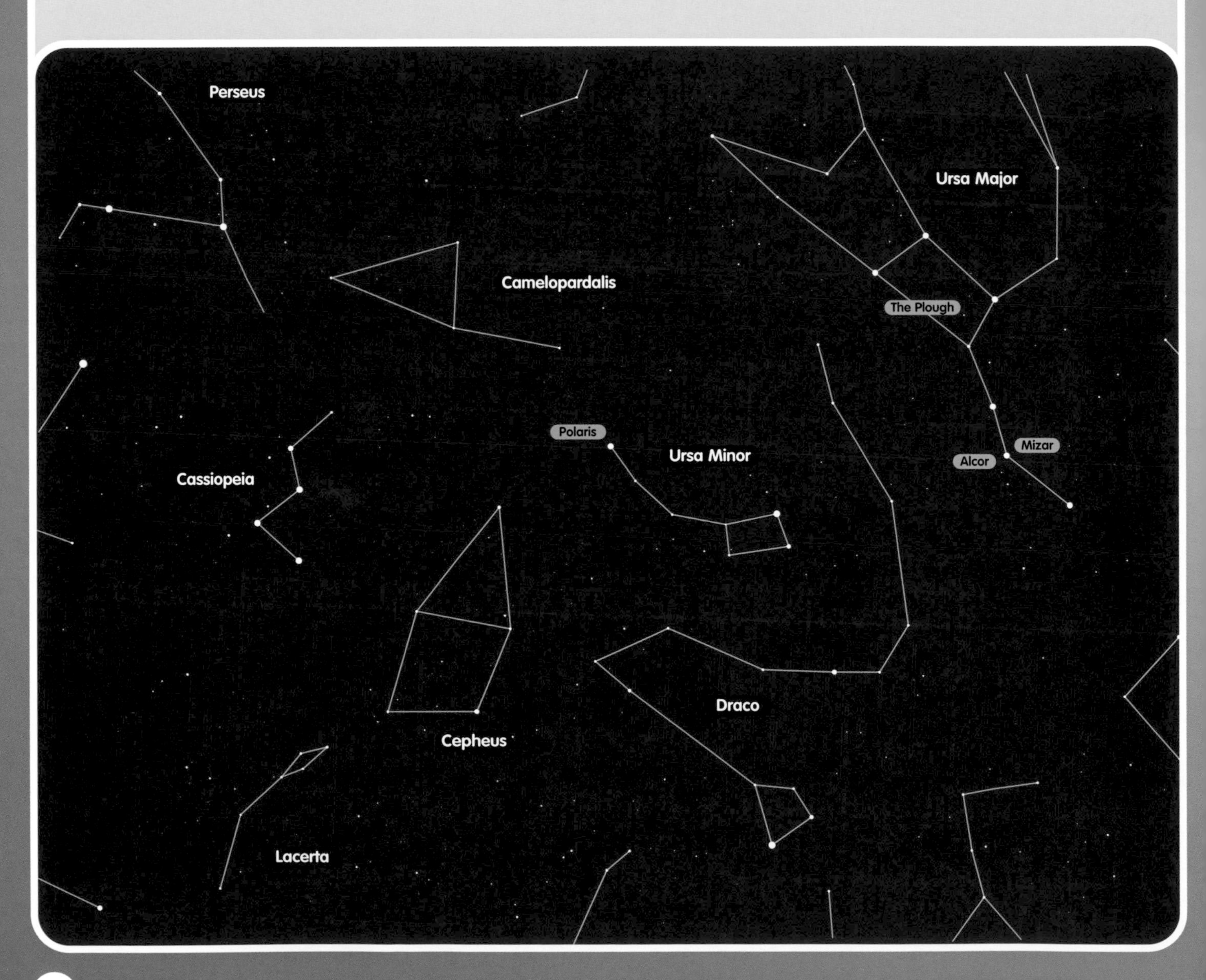

Observing Checklist

With your eyes only:

- Learn the shapes of the Plough, Cassiopeia, Cepheus and the Little Pan
- Find the Polaris, the North Star
- Find the double star Mizar & Alcor

With binoculars:

- Observe the double star Mizar & Alcor
- Find the seven stars in the Little Dipper, Ursa Minor
- Look around Cassiopeia and Perseus to find star clusters in the Milky Way

Circumpolar Constellations

The Earth spins on its axis every day, and our planet moves in orbit around the Sun once every year.

But one thing does not change: the Earth's North Pole always points in the same direction – to the North Star.

So the North Star, Polaris, is the only star that does not seem to move in our sky. It is always in the same place throughout the year.

Stars in the north move slowly around the North Star. Although these 'circumpolar' stars are not always in the same place, they can be seen at any time of year.

Skylights

The most famous all-year constellation is the Plough. It is shaped like a saucepan or a spoon, so it's sometimes called the Big Dipper. Three stars make the handle and four more make the pan or bowl of a spoon.

If you look at the middle star of the handle, you might see that it is actually two stars close together. The bright one is called Mizar and the faint one is Alcor, and together they make a double star. The pair of stars, sometimes called the Horse and Rider, were a test for good eyesight in the olden days. In 1995, the Hubble Space Telescope, collecting light for ten days, took a photograph of a tiny bit of sky near the Plough. This photograph, called the Hubble Deep Field, showed 3,000 galaxies far away, towards the edge of the universe.

The Plough is part of a bigger star group called the Great Bear or Ursa Major. The end two stars of the Plough point straight at the North Star. Polaris, the North Star, is not very bright and is only famous because it does not move. It has always been useful to travelers because it shows where north is at night.

Dim stars lead away from Polaris towards the Plough, making the shape of a little saucepan: the Little Dipper. The little pan belongs to the Little Bear, Ursa Minor.

Between the two saucepans, a straggly line of stars makes the tail of Draco the Dragon and, if you follow this wiggly line far enough, you will find Draco's head in a ring of stars. Draco was the hero of the film 'Dragonheart'.

On the opposite side of Polaris to the Plough is a letter 'W' made from five stars. The 'W' is the throne of Queen Cassiopeia. Cassiopeia is high in the sky in fall/autumn and the Plough is high in the sky in spring.

Close to the queen is her husband, King Cepheus. His stars make a square and a triangle joined together. The stars of Cepheus are dim.

Polaris, the North Star, is not very bright and is only famous because it does not move.

Southern Hemisphere

Summer Sky Window

December to February

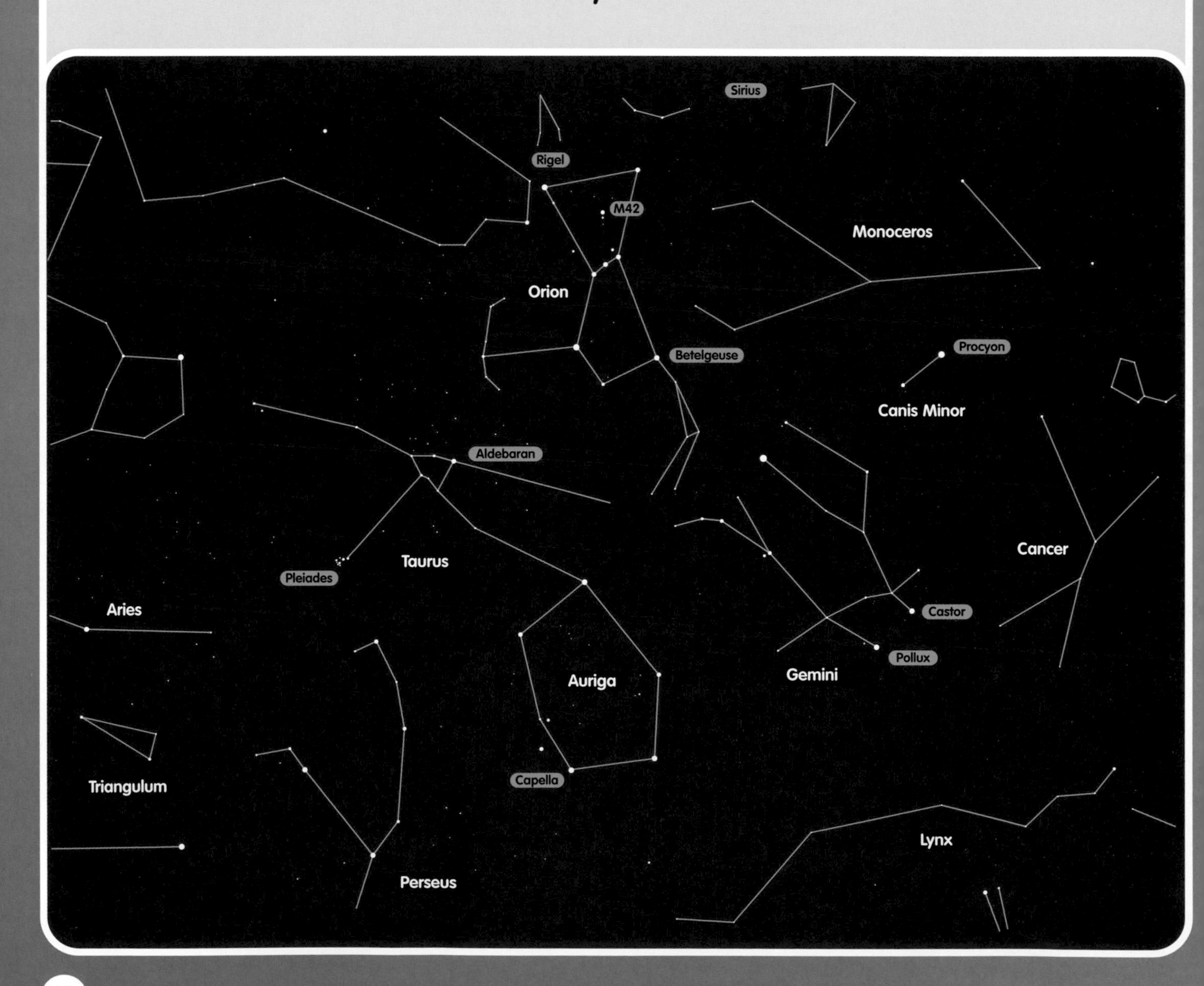

The maps for the Southern Hemisphere show the night sky as seen by people who live South of the equator. So if you live in Australia for example, use these charts.

Using the Sky Window

To find north, remember where the Sun has set, towards the west. Keep the sunset on your left and look ahead, and you will be looking north.

Summer nights are short, so you will have to stay up late to see the stars – but it's worth it!

Skylights

Orion the Hunter is our guide to the sky. First, find three stars in a straight line – these make Orion's belt. Orion should be an acrobat, as he stands upside down. Above the belt, two bright stars make Orion's knees and below the belt two more stars make his shoulders.

The star at the top is Rigel, a hot blue superstar. Rigel shines 60,000 times brighter than our Sun, at a distance of 800 light years.

The bottom right star is Betelgeuse, a red supergiant star. Betelgeuse is 40,000 times brighter than our Sun and 800 times wider. Some time in the future, Betelgeuse will explode as a supernova and leave a black hole in space. Luckily for us, this dangerous star is a safe 420 light years away.

Above Orion's belt, you may see a misty glimmer of light – a huge cloud of glowing gas called the Orion Nebula, code number M42. Inside this gas cloud, new stars are now being born. You can see the nebula clearly with binoculars or a telescope.

Follow the line of the belt up to the right. The bright star high in the sky is Sirius, the dog star, that appers to be the brightest star in our night sky. It is not really as bright as many other stars, but is closer, only eight light years away.

Look below Sirius and the next bright star is a puppy, Procyon. Go left from Procyon and you are back to Betelgeuse. Sirius, Procyon and Betelgeuse make a neat triangle of summer stars.

Below Procyon, you will find the twin stars of Gemini: the upper twin is Castor and the lower star is Pollux.

Gemini is the place for excitement on December 13 as there will be lots of shooting stars – the Geminid meteor shower, which comes every year.

Now go back to Orion's belt and follow the line of the belt down to the left. The orange star you see is Aldebaran, eye of Taurus the Bull. Keep going left until you see a fuzzy patch of stars. Now you are looking at a star cluster called the Pleiades, the Seven Sisters.

Low in the sky, the yellow constellation Capella, belonging to the star sign Auriga the Chariot Driver. Capella's light comes from two stars so close together that they appear as one.

Observing Checklist

With your eyes only:

- Learn the shapes of Orion, Taurus, Gemini and Capella
- Learn the stars Betelgeuse, Rigel, Sirius, Procyon, Aldebaran, Castor, Pollux and Capella
- Count how many stars you can see in the Seven Sisters cluster

With binoculars:

- Observe the colors of these stars – Betelgeuse, Rigel, Sirius, Procyon, Aldebaran, Castor, Pollux and Capella

To find out more:

- Look on pages 74 and 75 for Constellation Close-Ups of Orion and Taurus

Fall/Autumn Sky Window

March to May

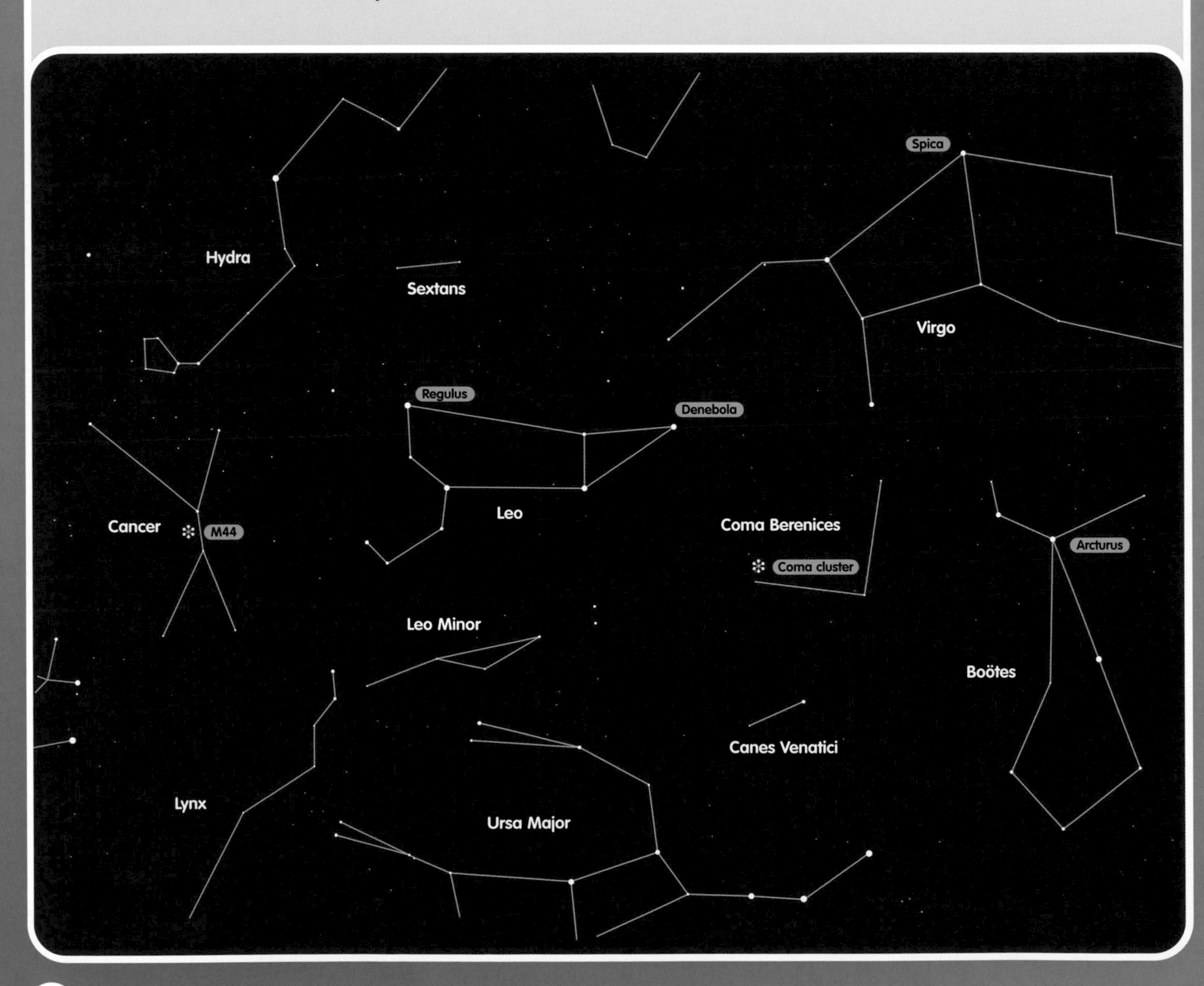

Observing Checklist

With your eyes only:

- Learn the shapes of Leo, Virgo and Boötes
- Learn the stars Regulus, Arcturus and Spica
- Find the Coma Berenices star cluster

With binoculars:

- Observe the colors of these stars – Regulus, Arcturus and Spica
- Observe the Coma Berenices star cluster
- Find the Beehive star cluster, M44

To find out more:

- Look on pages 69 and 70 for Constellation Close-Ups on Leo and Boötes

How to Use the Sky Window

Remember where the sun has set, roughly in the west. Keep the sunset on your left, look ahead and you will be looking north.

Skylights

Leo the Lion is hanging upside down in the south and to find Leo you need to look for stars making a big hook in the sky. The bright star at the top of the hook is Regulus, whose name means 'King Star' because it is near the path of planets across the sky. Regulus is 150 times brighter than our Sun and 80 light years away. To the right of Regulus, other stars make the lion's body, and his tail is a star called Denebola.

Saturn will be near Leo in 2008 and 2009, so the lion will have an extra bright 'star'.

To the right of Leo, the bright red giant star is Arcturus, the fourth brightest star in the sky. Arcturus is a fast-moving star, traveling through the galaxy at 56 miles/90 kilometers per second – that's about 200,000 miles per hour/325,000 kilometers per hour.

Below Arcturus, a kite-shape of dimmer stars points to the ground. This is the shape of Boötes the Shepherd.

Above Arcturus is a blue star, high in the sky. This is the superstar Spica, brightest star in Virgo, 14,000 times brighter than the Sun and shining about 260 light years away.

Astronomers like Virgo because it's the place to look for galaxies – when we look at Virgo, we are looking out of the Milky Way. With big telescopes, hundreds of galaxies appear as tiny blobs of light. There are 1,500 galaxies in the Virgo cluster. Our own group of galaxies, including the Milky Way, belongs to the Virgo supercluster of galaxies.

By 2010 and 2011 Saturn will have moved to Virgo.

Between Leo and Boötes are the stars of Coma Berenices. In legend, these stars make the hair of Queen Berenices.

Although the stars are dim, they are collected in a beautiful cluster. To see them more clearly, look to the right or left – the cluster looks like a fountain of stars in the edge of your eye. This cluster looks good through binoculars.

Over to the left of Leo is Cancer the Crab, whose stars are in the shape of a dim letter 'Y'. Like Leo and Virgo, Cancer is one of the 12 signs of the Zodiac. The Zodiac constellations mark the track of the Sun, Moon and planets across the sky.

Just below the middle star of Cancer is a nice star cluster, which you might just see with only your eyes. This is called the Beehive, and its number is M44. With binoculars, and a bit of imagination, this star cluster looks like a swarm of bees.

The superstar Spica, the brightest star in Virgo, is 14,000 times brighter than the Sun.

Southern Hemisphere

Winter Sky Window

June to August

Observing Checklist

With your eyes only:

- Learn the shapes of Scorpius, Sagittarius, Ophiuchus and Hercules
- Learn the stars Antares, Vega and Altair
- Look for the cloudy band of the Milky Way

With binoculars:

- Observe the colors of these stars – Antares, Vega and Altair
- Find the star clusters M6 and M7
- Find the Lagoon Nebula, M8, and the Trifid Nebula, M20
- Look at the Milky Way to find star clusters

With a telescope:

- Observe the Lagoon Nebula and make a drawing of what you see

How to Use the Sky Window

To find north, remember the direction of sunset, which is roughly west. Keep the sunset on your left, look ahead and you will be looking north.

Skylights

The long nights of winter make it a good time to look at the night sky. The winter stars are wonderful, and you can find lots of objects in the Milky Way.

High in the sky is Scorpius the Scorpion who, in legend, stung Orion and killed the great hunter. The brightest star in the Scorpion is the red giant Antares, whose name – 'Ant-Ares' – means that it looks like Mars. It's a supergiant star, shining 10,000 times brighter than our Sun from a distance of 600 light years.

Above Antares, a curl of stars makes the sting in the scorpion's tail and, below, more stars make his claws.

To the right of Scorpius sits Sagittarius the Archer, who really looks like a teapot upside down! Four stars make the pot, two stars to the right make his handle and another to the left draws its spout.

When you look at Sagittarius, you are looking towards the middle of our galaxy, the Milky Way. You cannot see the middle because there is too much gas and dust in the way, but we can see millions of stars in the spiral arms.

You will find lots of great star clusters and nebulae in the Milky Way from Sagittarius down to the ground.

Below the teapot are two nebulae called the Lagoon (M8) and the Trifid (M20) – both fuzzy glows of gas in binoculars or a telescope. The Lagoon is one of the brightest gas clouds, while the Trifid is much dimmer.

Above these nebulae are two terrific star clusters: M7 (called Ptolemy's cluster) and, below and left, M6, the Butterfly cluster. Both M6 and M7 contain about 130 stars and are best seen through binoculars.

Below Scorpius, a huge faint diamond of stars makes Ophiuchus (pronounced Off-ee-oo-kus), which carries a snake called Serpens. Stars to his right are the snake's body, Serpens Cauda, and those to the left are his head, Serpens Caput.

During 2008 and 2009, Jupiter will be high in the sky. This will be a great chance to observe the giant planet and its four great moons.

When you look at Sagittarius, you're looking towards the middle of the Milky Way. We can't see the middle because there is gas and dust in the way.

Southern Hemisphere

Spring Sky Window

September to November

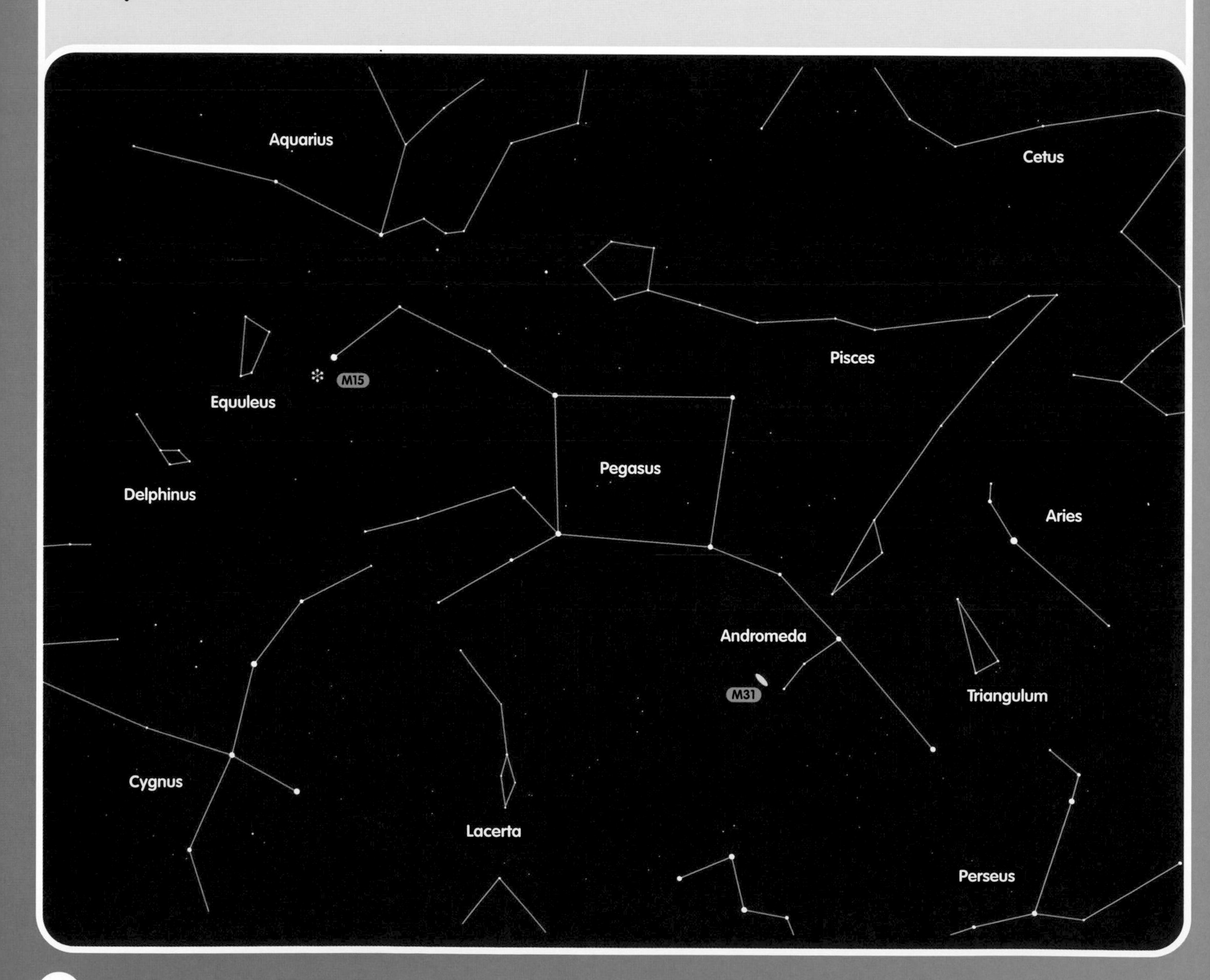

Observing Checklist

With your eyes only:

- Learn the shapes of Pegasus, Andromeda, Cetus, Aquarius, Pisces and Delphinus
- Find the Andromeda Galaxy, M31

With binoculars:

- Observe the Andromeda Galaxy, M31

With a telescope:

- Observe the Andromeda Galaxy, M31 and make a drawing
- Observe the globular cluster M15 and make a drawing

To find out more:

- Look on pages 78 and 79 for Constellation Close-Ups on Pegasus and Andromeda

How to Use the Sky Window

To find north, first notice where the sun sets, in the west. Keep the sunset on your left and look in front towards the north.

Skylights

Looking north, you will see a big square made of four stars. This is the Square of Pegasus, belonging to the famous flying horse. The square is the horse's body and, from top left, a line of stars goes upwards then bends down to make his head and neck. From bottom left, more stars make his legs.

Just in front of Pegasus there is a superstar cluster with the code number M15, which can be seen with a telescope. First follow the stars along the horse's neck and down his head, then follow the same line to a fuzzy star. A telescope will show a grainy star cluster. This is a globular cluster, made of thousands of stars packed together in a ball.

From bottom right of the square, a line of four stars makes Andromeda the Princess. To the left of Andromeda, you may see a hazy glow of light – this is the furthest thing you can see using your eyes. The glow comes from the Andromeda galaxy, M31, which is 2.5 million light years away. The light from this galaxy has taken over two million years to reach your eyes.

In the old story, Andromeda was rescued from a sea monster, which still lurks in the sky. Look above to the right of the square and there is the monster, Cetus. Dim stars make his body, then down to the right stretches his long neck and round head. The monster floats upside down after his fight with Perseus.

This is a watery part of the sky. Above Pegasus we find Pisces the Fish and Aquarius the Water Carrier. To the left of the square, Delphinus the Dolphin is playing in the sea. Look for a tiny diamond of stars with another above. The diamond is the dolphin's body and the top star makes his tail. The dolphin is a lovely little constellation to observe with binoculars.

None of the watery stars are very bright, but a brilliant 'star' will drift across their sky between 2010 and 2012. The star is actually the giant planet Jupiter, which you will be able to observe and see the movement of its four big moons.

Meteors, which belong to the Perseid meteor shower, are often very bright and sometimes leave trails of light that last for several minutes.

Southern Hemisphere

Sky Window

All-year Stars

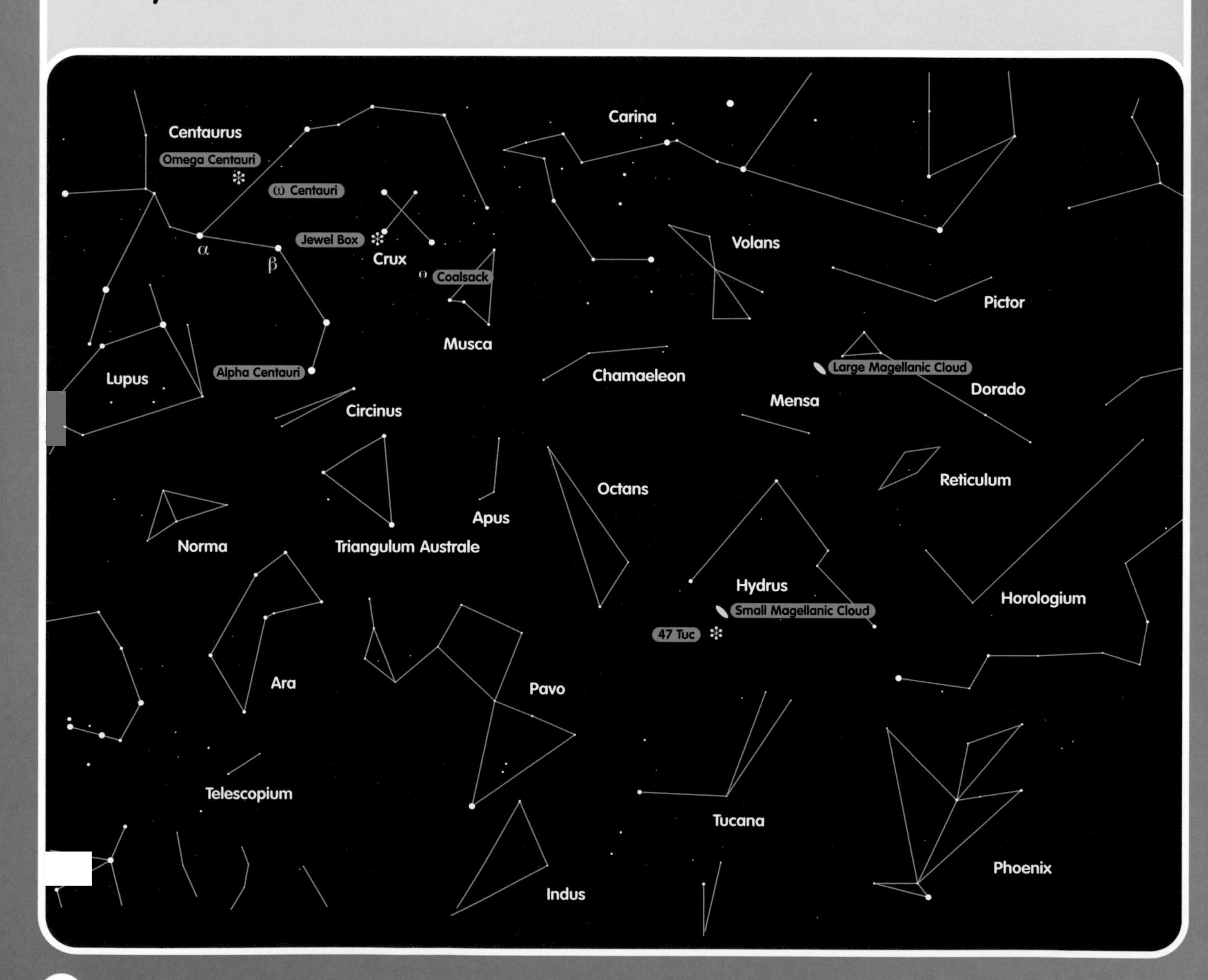

Observing Checklist

With your eyes only:

- Find the Southern Cross, the pointers, Alpha Centauri, the Small Magellanic Cloud and the large Magellanic Cloud

With binoculars or telescope, observe:

- The colors of the four stars in the Southern Cross
- The Jewel Box star cluster
- The Coalsack dark nebula
- Omega Centauri globular cluster
- 47 Tucanae globular cluster
- The Large Magellanic Cloud
- The Small Magellanic Cloud

Circumpolar Constellations

The Earth spins on its axis and moves in orbit around the Sun, but the Earth's South Pole always points to the same place in the sky. This point is called the South Celestial Pole and it is always in the same place in the sky.

Sadly, there is no bright star to mark the south pole of the sky.

Stars near the Celestial Pole move slowly around it during a single night and over the year.

Although they are not always in the same place, these stars – called the circumpolar stars – can be seen at any time of year.

Skylights

The most famous constellation is Crux, the Southern Cross. It is the smallest constellation but one of the most beautiful, and is high in the evening sky between April and June. Four stars make the cross, and binoculars will show that one of them is a different color to the other three.

Close to the left arm of the cross is an amazing star cluster called the Jewel Box, which looks wonderful through binoculars or a telescope. It has hot blue stars mixed with red giants, all about 8,000 light years away.

The sky seems to darken near the Jewel Box. Here a thick, dark cloud of dust – the Coalsack nebula – hides the stars.

A little further from the cross is the best globular cluster in the whole sky: Omega Centauri. At first it looks like a fuzzy star, but binoculars will show that it is a grainy ball, and a telescope will show lots of stars packed together. Omega Centauri contains millions of old stars and is about 18,000 light years away.

To the left of the Southern Cross are two bright stars called 'the pointers'. The second pointer, furthest from the cross, is Alpha Centauri, the closest star to our Sun. Actually, Alpha Centauri is made of three stars close together, the nearest being a small red dwarf star called Proxima Centauri.

Alpha Centauri's stars are 4.2 light years away. Even with the fastest rocket, it would take 100,000 years to travel there!

On a clear dark night, you might see two wispy star clouds, looking like pieces of cotton wool: the Magellanic Clouds. Each one is a dwarf galaxy made of millions of stars. The two galaxies – 160,000 light years from the Milky Way and about 75,000 light years apart – are best seen from November through to March.

Close to the Small Magellanic Cloud (SMC) is another treasure: 47 Tucanae. Usually called '47 Tuc', it appears as a fuzzy star if you look at it with your eyes, but a telescope will show lots of stars packed together. This is a superb globular cluster, a ball of thousands of stars about 13,000 light years away.

Alpha Centauri's stars are 4.2 light years away. The fastest rocket would take 100,000 years to get there!

Constellation Close-Up

Orion the Hunter

The Legend

Orion, a great hunter, holds a club in one hand and a shield in the other. Orion's adventures ended when he was killed by a scorpion.

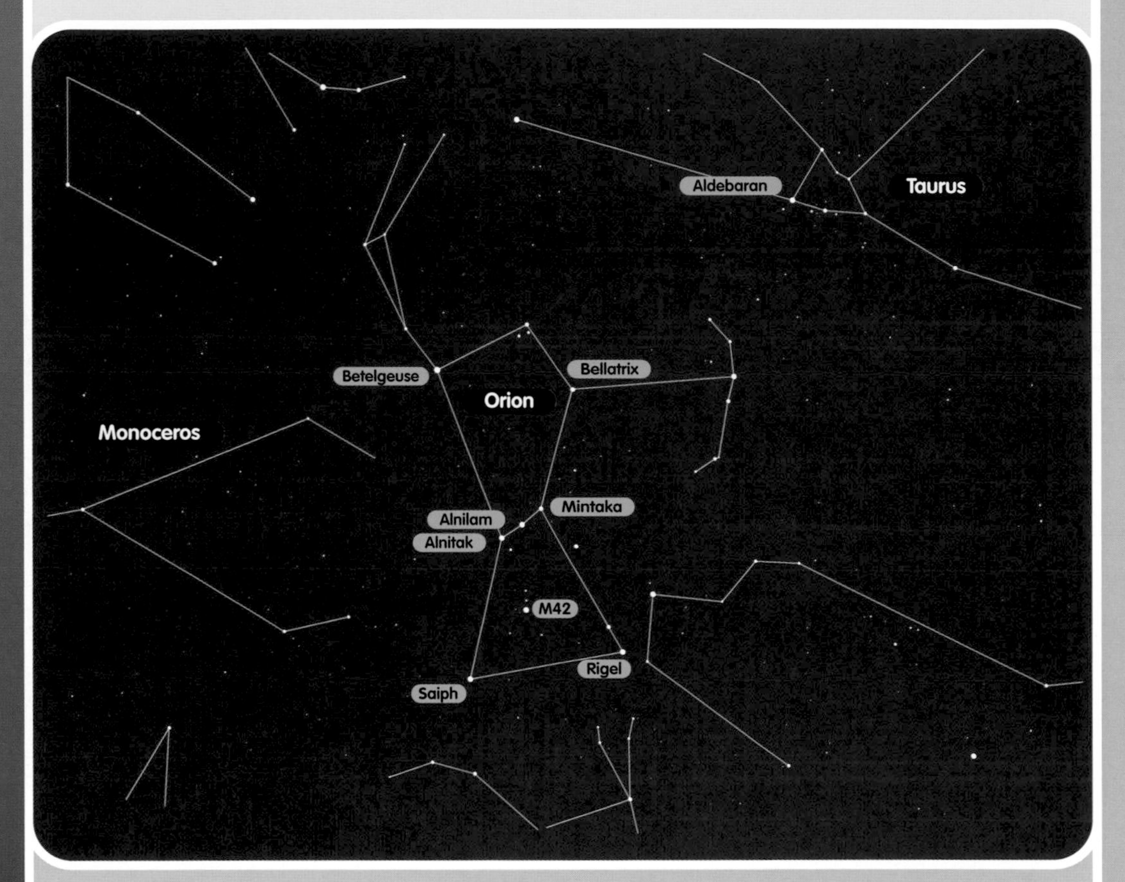

Observing Orion

With your eyes, trace the outline of Orion. Three stars in a line – Alnitak, Alnilam and Mintaka – make his belt.

Above the belt, Orion's shoulders are made by Betelgeuse and Bellatrix and, below the belt, his knees are Saiph and Rigel.

Above Betelgeuse, find the dim stars that make Orion's club. To the right of Bellatrix, look for the line of stars in his shield.

Below Orion's belt hangs his sword, with the tip a hazy glow of light. This is the Great Orion Nebula, code number M42.

Observation Checklist

Here are some great observations you can do with binoculars or a telescope.

Star colors:
Alnitak, Alnilam, Mintaka, Betelgeuse, Bellatrix, Saiph, Mintaka.

Nebula:
Look at the Orion Nebula, a cloud of gas and dust, and the best nebula you can see. Using high power on your telescope you might see the tiny group of four stars that makes the nebula glow.

Taurus the Bull

The Legend

The great god Jupiter came to Earth in many disguises, and one of these was Taurus, a white bull. In the sky, we see the bull's head and horns made of stars.

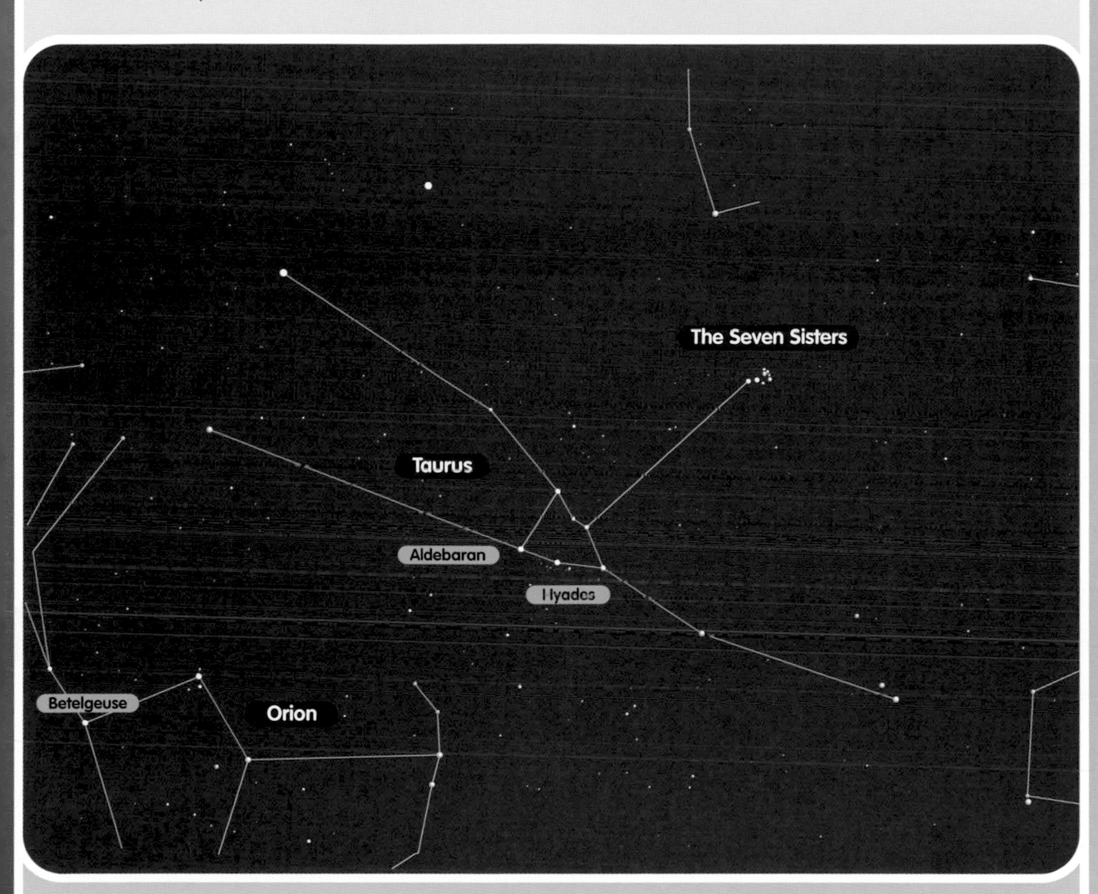

Observing Taurus

First find the orange eye of the bull: a red giant star called Aldebaran. Around Aldebaran, more stars – the Hyades star cluster – make the bull's nose.

To the left of the bull's eye, two dimmer stars make his horns.

Above Aldebaran is another star cluster called the Pleiades, code number M45. In legend, the Pleiades were seven beautiful sisters.

Some people can see only five or six stars, while others claim to see 12 or 13. To see the Pleiades, do not look straight at them; instead, look slightly to the side, for the edge of your eyes sees stars more clearly.

Observation Checklist

With your eyes:
Count the stars in the Pleiades.

With binoculars or a telescope:
Look at the Pleiades star cluster, the Seven Sisters. There are many more than seven, and this is the best cluster in the sky.

Look at the stars just to the right and above Aldebaran. Binoculars are good for viewing this.

These stars make the Hyades star cluster and they shine with many colors. There are lots of double stars too.

Constellation Close-Up

Leo the Lion

The Legend

Leo was a lion whose skin could not be pierced by any weapon. Hercules managed to kill the fearsome lion by choking him to death.

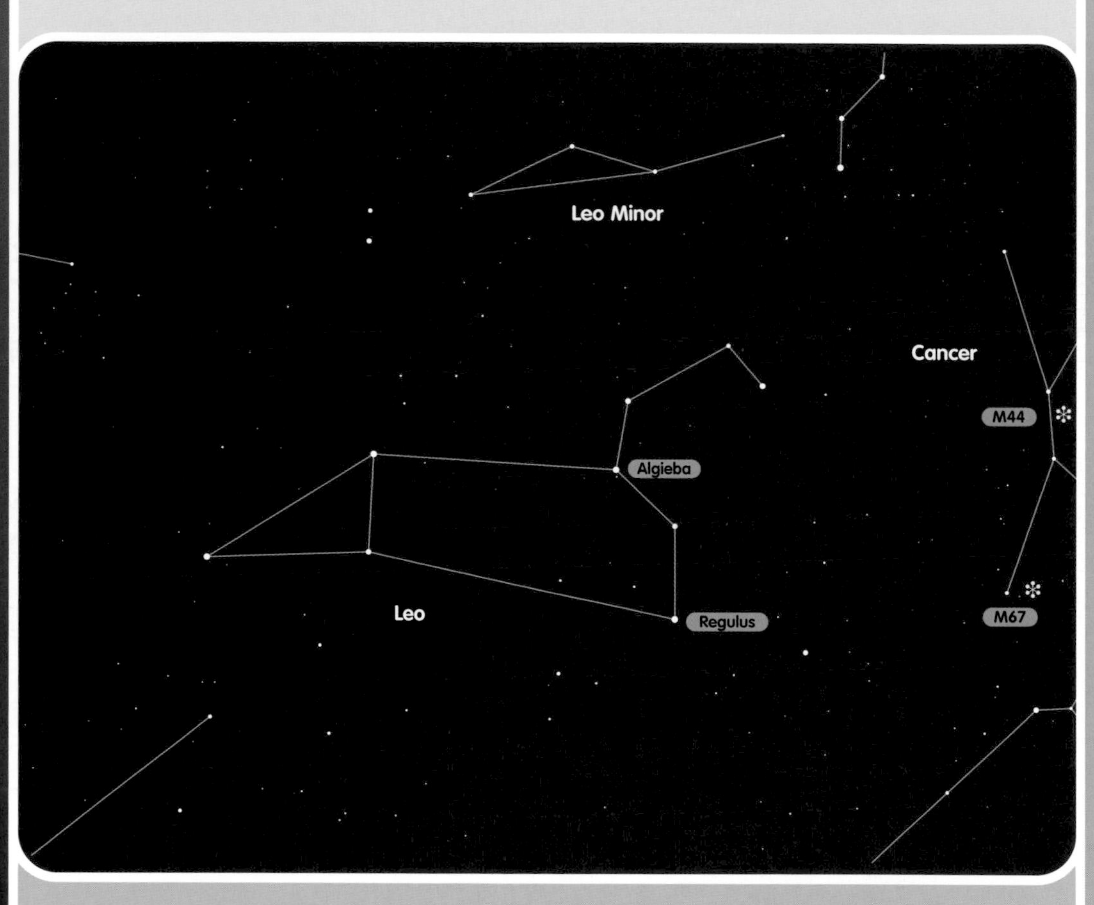

Observing Leo

Leo's head is made from a hook of stars, looking like a question mark drawn backwards. The bright star at the bottom of the hook is Regulus and the star at the bend of Leo's neck is called Algieba, a lovely double star. More stars to the left make the lion's body and tail. To the right of Leo are the dim stars of Cancer the Crab. If you can see them, they make an upside down letter 'Y'. There are two good star clusters in Cancer . The brightest, in the middle of Cancer, is M44, the Beehive cluster, whose stars look like a swarm of bees. Another star cluster, M67, is near the bottom of Cancer.

Observation Checklist

Star clusters in Cancer look best through binoculars.

Star clusters:
Observe M44, the Beehive.
Observe M67.

Double star:
Find Algieba in a telescope. Look for two stars close together.

Boötes the Shepherd

The Legend

Some legends say Boötes was the farmer who used the stars of the Plough. Others say he was a shepherd guarding the stars from the hungry Great Bear.

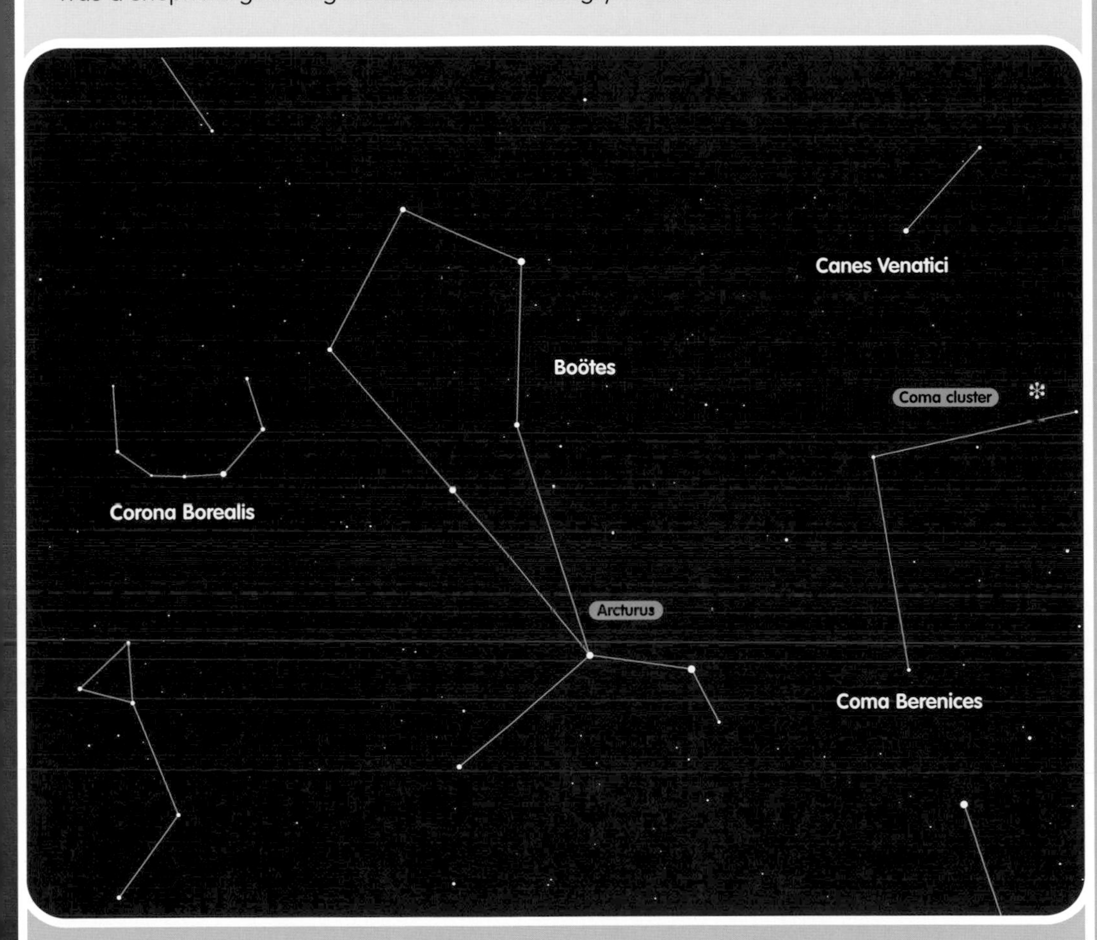

Observing Boötes

The brightest star in Boötes is a red giant called Arcturus. Arcturus means the 'bear guard'. This is an unusual star, which does not move around the galaxy like the Sun and most stars. Instead, it is moving through the galaxy from above to below. To the left of Boötes, you may see stars in a curve. This is the Northern Crown, Corona Borealis.

To the right of Boötes is a big star cluster in the constellation Coma Berenices.

Observation Checklist

Look at Arcturus with your eyes only, then use binoculars or a telescope.

With your eyes and binoculars, find the Northern Crown, Corona Borealis.

With your eyes, find the star cluster of Coma Berenices. It will help to look slightly to the side, for the edges of your eyes make dim stars seem brighter. Then look at the Coma cluster with binoculars to make the stars brighter still.

Pegasus the Flying Horse

The Legend

When Perseus killed the Medusa, he chopped off her head. From the blood sprang Pegasus, a flying horse, who then helped Perseus to rescue Princess Andromeda from the sea monster.

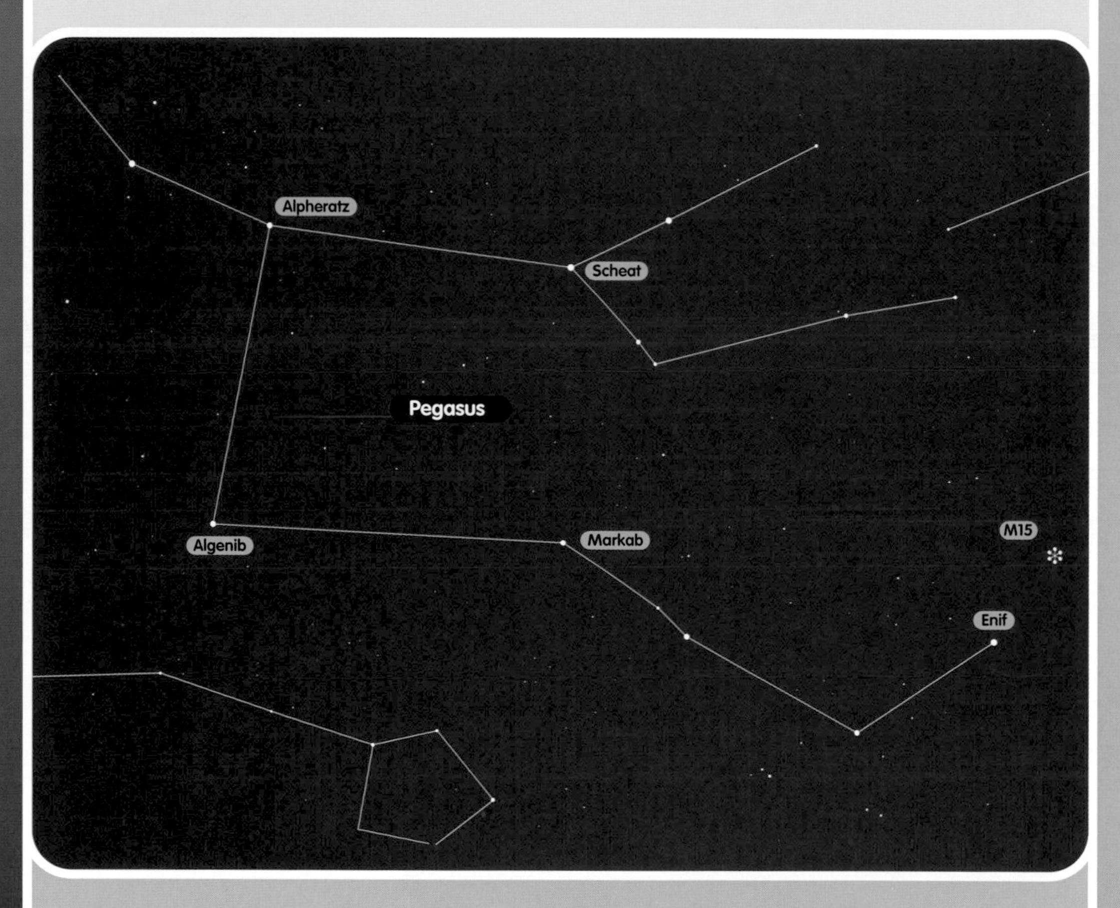

Observing Pegasus

The body of Pegasus is a big square of stars. From the bottom right of the square, more stars make the horse's neck and bend up for his head. Pegasus is upside down. The star at the horse's nose is called Enif. From the top right, two lines of stars make the front legs of Pegasus, but the rest of the horse cannot be seen – he is leaping from the sea after his fight with the monster.

Pegasus can show you how good your night sky is: the more stars you can see inside the square, the better your view. If you can see only three or four stars, the sky is poor, while five to ten stars means the sky is OK.

Observation Checklist

With your eyes:
Count the stars inside the square. Rate your sky poor, OK or good.

With binoculars:
Observe the color of the stars Alpheratz, Algenib, Markab, Scheat and Enif.

With a telescope:
Look at the sky just beyond Enif and you will find a grainy ball of stars. This is a globular cluster, code named M15.
M15 is an old cluster, about 13 billion years old. It is 34,000 light years away and has thousands of stars.

Andromeda the Princess

The Legend

Andromeda was a princess whose mother, Queen Cassiopeia, annoyed the god of the sea. The sea-god sent a sea monster to terrify her people. To give the monster a meal and make him go away, Andromeda was chained to a rock on the beach. Luckily, the princess was rescued by Perseus.

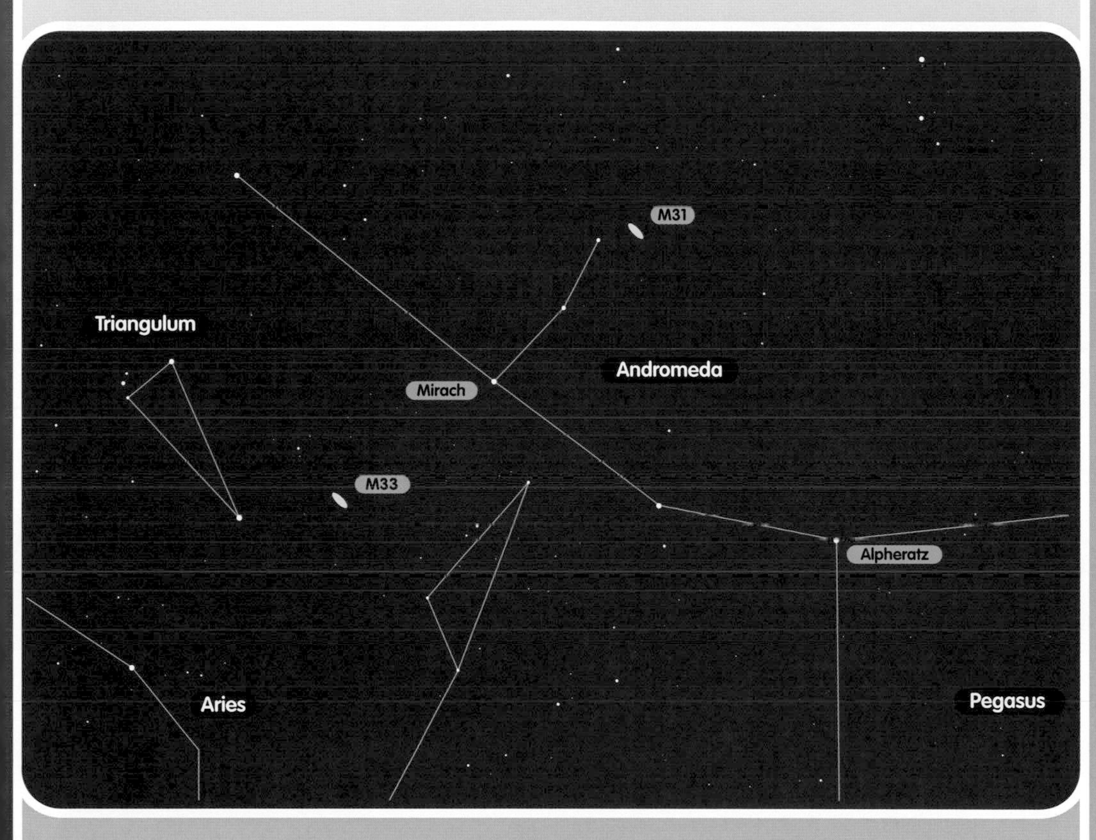

Observing Andromeda

Andromeda is made from two lines of stars stretching out from the top left of the Square of Pegasus. Alpheratz is the star at Andromeda's head. Andromeda is famous as the home of the Great Galaxy.

To find the galaxy, count three stars from the square. Stop at Mirach, then count three stars upwards. Close to the last star is a fuzzy glow in the sky. This is the Andromeda Galaxy, code number M31.

The galaxy, the nearest big spiral galaxy to the Milky Way, is 2.5 million light years away.

Observation Checklist

With your eyes and binoculars:
Find the Andromeda galaxy, M31. It looks like a misty glow.

With a telescope:
Observe the Andromeda galaxy. You can only really see the center of the galaxy, where it is brightest. If you could see all the spiral arms, it would look as big as three Full Moons.
You could try to find another galaxy, M33, below Mirach at about the same distance away as M31. But this galaxy is very hard to see.

Observing the Moon

The Moon is the best place to start observing the sky, because it's very bright and looks different every night. Our guide shows what to look for.

With your eyes:
Look at the Moon every clear night – try to do this for a month – and draw the shape of the Moon in your observing book each night. This way you can follow the Moon as it moves around the Earth and make a record of the phases of the Moon.

With binoculars:
Binoculars magnify the Moon so that you can see the dark maria and some round craters. Next Full Moon, see if you can spot the rabbit and the lady's face.

With a telescope:
A telescope magnifies the Moon so that you can see the dark maria and lots of craters.
From crescent to gibbous is the best time to look at craters and mountains.

Craters stand out best where the bright day meets the dark night.

From gibbous to Full Moon is the best time to look at the dark maria and bright rays coming from craters.

If you want to find out more about the Moon, please turn to pages 14 and 15.

Observing Mercury

Planet Guide

Planets can be found among the stars using your eyes, or through binoculars – where they will appear brighter, or you can magnify your viewing through a telescope. Our Planet Guide gives you tips on when to look for planets and where to find them in the sky.

2008: May 14
2009: April 26
2010: April 8
2011: March 23
2012: March 5

Mercury is difficult to find because it is always close to the Sun. The best time to find it is when it stays in the sky just after sunset. The dates on the left show the best times to see this little planet – you will find Mercury a few days before and after these dates.

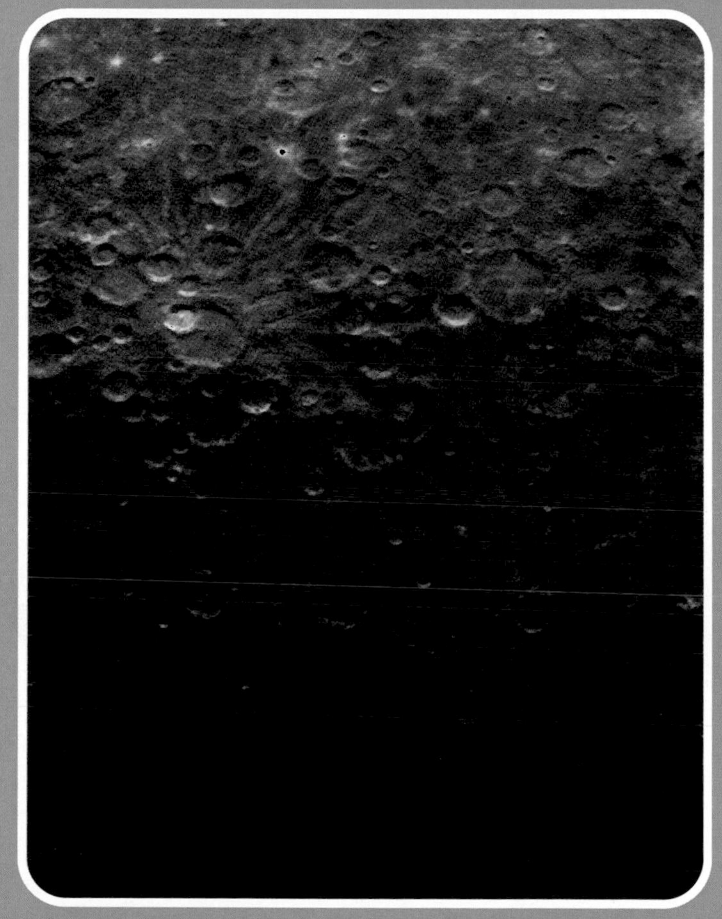

With your eyes:
Look towards the west just after sunset, in the direction of the setting Sun. Mercury will look like a steady bright star.

With binoculars:
Mercury will become a bright dot.

With a telescope:
You will see Mercury shows a phase like the Moon. It might look like a tiny crescent or half Moon. Even with the biggest telescope in the world, you cannot see the craters.

DANGER: you must only look for Mercury after the Sun has disappeared. If you accidentally see the Sun through binoculars or a telescope, it will damage your eyes.

Observing Venus

Venus is brighter than any star, so it's easy to find. When you look at Venus you cannot see the surface, for it is always covered in clouds. You see sunlight reflected from the clouds. Because Venus is closer to the Sun than we are, it shows phases like the Moon. Unlike the Moon, however, Venus is brighter when it is a crescent. That's because it is nearer to Earth.

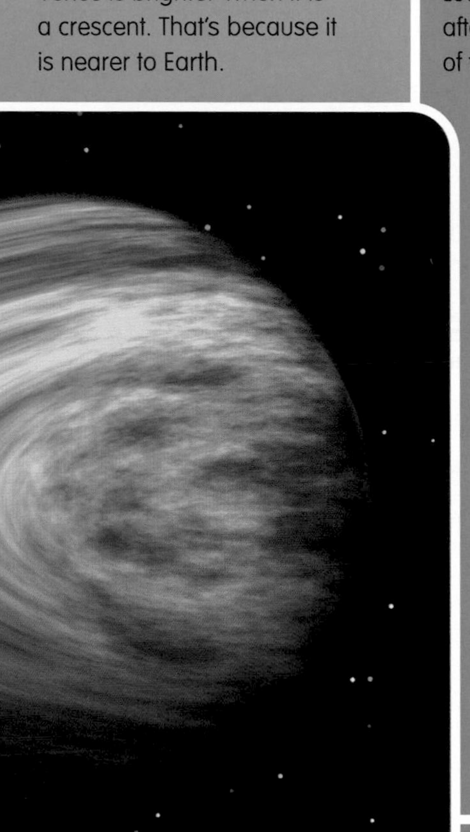

Observing

With your eyes:
Look towards the west just after sunset, in the direction of the setting Sun. Venus will shine like a brilliant star, looking much brighter than any of the stars. You can see it easily before the sky goes dark.

With binoculars:
Venus will look like a very bright dot. By looking very hard, you may just see a phase like the Moon.

With a telescope:
Venus looks very bright! Using high power may help to cut down the light. If you look carefully, you may see that Venus shows a phase like a tiny Moon. Try to watch the planet for several weeks – then you will see the phase change.

DANGER: you must only look for Venus after the Sun has disappeared. If you accidentally see the Sun through binoculars or a telescope, it will damage your eyes.

Observing Dates

The dates here are the best times to look; you can see Venus a few weeks before and after these dates. It is fairly close to the Sun. The best time to see Venus is just after sunset.

2009: January 14

2010: August 20

2011: January 8

2012: March 27

Observing Mars

Mars glows with an orange-red light, looking like a red giant star – but it does not twinkle.

With a telescope you can see markings on its surface. The view of Mars can be spoiled by turbulence in the Earth's atmosphere, which is known as wobbly air.

When you look at Mars through a telescope, you must be patient – wait until our air settles and the markings will become clear.

Observing Dates

Mars comes close to Earth every two years. The dates below tell you the best times to see the planet and where you can find it. Mars will be high in the sky for the next few years.

2007 – 2008:
November 2007 to March 2008.
Closest: December 24
Constellation: Gemini

2010:
January to April
Best: January 29
Constellation: Cancer

2012:
February to May
Best: March 3
Constellation: Leo

2014:
March to May
Best: April 8
Constellation: Virgo

Observing

With your eyes:
Mars glows a bright orange-red color. If you watch the planet for several weeks, you may notice it has moved across the stars. The word 'planet' means a 'wandering star'.

With binoculars:
The color becomes brighter and Mars is a tiny dot.

With a telescope:
Mars is an orange circle, and the color comes from the planet's red deserts. You may see dark marks, to these are areas of rock where the sand has blown away. The largest dark mark is called Syrtis Major. At the upper or lower edge of Mars, you may see a lighter mark – an ice cap near one of the poles.

Observing Jupiter

Jupiter, a large, bright planet with four big moons and countless little ones, is the best planet to look at with binoculars and telescope.

Observing

With your eyes:
Jupiter looks like a yellowish star, shining steadily. It does not twinkle like the stars.

With binoculars:
Jupiter is a small dot, a tiny circle. You will see the moons, looking like little stars in a row. Sometimes all four moons are on one side of the planet; on other nights you will see them on either side.

If you make drawings on several nights, put them together and you'll see how they have moved – doing what Galileo did 400 years ago. Your binoculars will be better than Galileo's telescope, though!

With a telescope:
Jupiter is a small circle. On medium or high power you can see dark stripes and white stripes across it. Watch carefully for a darker oval shape in the stripes – that is the Great Red Spot. On low power, the planet looks small but you can see the moons.

Observing Dates

2008:
July to August
Best: July 9
Constellation: Sagittarius
Note: In 2008, Jupiter is very low in the south. In July, the sky stays light late into evening.

2009:
August to September
Best: August 14
Constellation: Capricorn
Note: In 2009, Jupiter is low in the south

2010:
September to November
Best: September 21
Constellation: Pisces
Note: Jupiter is higher in the south and the nights are longer.

2011:
October to December
Best: October 29
Constellation: Pisces
Note: Jupiter is much higher in the south in dark skies.

2012 – 2013:
November to March 2013
Best: December 3
Constellation: Taurus
Note: A great time for observing Jupiter as it's high in a dark winter sky.

Observing Saturn

Saturn is the most beautiful planet to see in a telescope, as any scope will show its rings. You may also see some of its moons, particularly Titan, looking like tiny stars around the planet.

Observing

With your eyes:
Saturn looks like a yellow star, shining with a still light. It does not twinkle like the stars.

With binoculars:
Saturn looks like a tiny yellow dot. Although you cannot see the rings, you may notice that it is not quite round.

With a telescope:
Saturn is a small circle and you can see the rings. Use the highest power you can and look on the planet to see if there are faint stripes around it.

Now look at the rings. You might just see a dark line running through them, best seen at the sides of the rings. The dark mark is the Cassini Division, a gap in the rings. You might also see that the outside ring is darker than the ring inside it.

As you look longer, Saturn appears to hang in space. It will look 3-D.

Observing Dates

2008:
February to April
Best: February 24
Constellation: Leo

2009:
February to May
Best: March 8
Constellation: Leo

2010:
March to May
Best March 22
Constellation: Virgo

2011:
March to June
Best: April 4
Constellation: Virgo

2012:
March to June
Best: April 15
Constellation: Virgo

Naked-Eye Astronomy

Eyes on the Universe

You don't need to buy expensive equipment to observe the night sky – your eyes and a clear sky are all you need. In fact, when you start observing, it's a good idea to get to know the sky by using just your eyes.

Averted Vision

Astronomers use a great trick for viewing things better just using their eyes, and you can learn it in a few seconds. You might be looking at a dim object in the sky, like a star cluster. When you look straight at it, the stars appear dim. But if you look slightly to the side of the cluster it looks brighter and you can see it better. The trick, which is called averted vision, works because the edge of each eye sees dim light better than the middle.

The most important thing for any observer who's starting out is to learn the constellations. Get to know their names and patterns and the names of their bright stars, and work out how one star sign leads to another. Use our star maps to help, and soon you will start to recognize the important star patterns in the sky.

The Moon is a good starting point. As the Moon orbits the Earth, it reflects different amounts of sunlight and the phase changes. You may want to draw the Moon each clear night and see how it changes over several weeks. Look at our Moon phase illustration on pages 14 and 15 and work out where the Moon is in its orbit.

But shooting stars make the most exciting naked-eye project. Telescopes are not good for seeing meteors because they move so quickly across the sky. All you need is a sun-lounger, a watch, a paper and pencil, and a red-light flashlight. Choose a night when there is a meteor shower – August 12 and December 13 are best. Remember to always ask permission from an adult before venturing out at night.

Telescopes are not good for seeing satellites – your eyes are best. Although satellites are small, about car-size usually, you can see them easily. They are shiny and reflect lots of sunlight from their orbit a few hundred miles high. It's

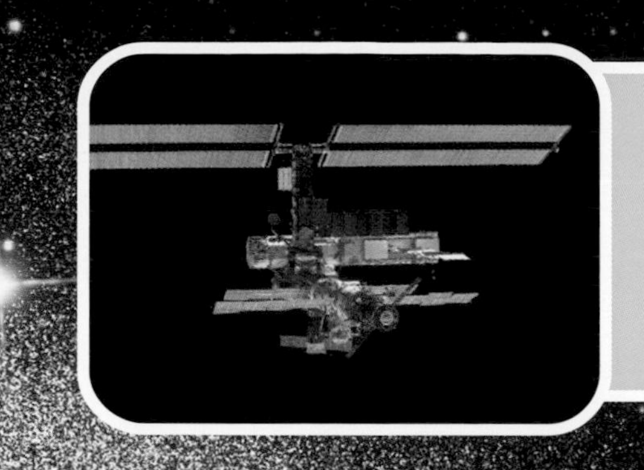

Heavens-Above

The most exciting satellite to look for is the International Space Station, while the brightest are the Iridium satellites that flash brighter than any star just for a few seconds.

easy to see the difference between satellites, airplanes and shooting stars. Satellites move slowly across the sky in about five minutes, while shooting stars flash across in a couple of seconds. Airplanes have flashing lights; satellites have a steady light.

Any night, while you are stargazing, be ready for the unexpected. You may see a super-bright meteor, a fireball, flash through the atmosphere. You may even see the red-green glow of an aurora. It's fun, it's free and you only need your eyes.

Observing with Binoculars

You can see lots in the night sky using only your eyes, but eventually you will want to see more. The next step is to get a pair of binoculars. You may have some in your house already.

Using a pair of binoculars is like having enormous eyes. Our eyes are small and collect only a small amount of light, but binoculars collect lots of light so when you look through them everything looks bigger and brighter.

Binoculars are great for astronomy and in some ways are better than a telescope. They are certainly easier to use. A telescope can be big and heavy and take a long time to set up. With binoculars, you just put the strap around your neck and walk outside – and you are ready to observe.

Binoculars are more comfortable to use because you use both eyes. And with binoculars you can see more of the sky – it's called a wide field of view – so you can find stars and planets more quickly. Binoculars are not just useful for astronomy, you can use them for down-to-earth viewing too. Most important though, they are much cheaper than a telescope.

Binoculars have two pairs of lenses. The big lenses, facing the object, are called the objectives. The small lenses, the ones you put next to your eye, are the eyepieces.

Most binoculars have two numbers written on them somewhere. The numbers may be something like 8 x 30 or 10 x 50 and they tell you what the binoculars will do.

The first number is the magnification, so 8 means they magnify eight times, 10 means ten times. The second number is the size of the big lens, the objective, in millimeters. So 30 means a 30mm lens and 50 is a 50mm lens. The bigger the number, the more light the binoculars collect – but the heavier the binoculars become.

Danger from the Sun

Never try to use binoculars to look at the Sun. Binoculars will magnify the heat and light from the Sun and focus it all on the inside of your eyes. It will burn the back of your eyes and you could go blind.

Double Vision

For astronomy, 10 x 50 binoculars are fine. They magnify, collect plenty of light and can be carried easily.

So take care of your binoculars. After you have used them, put the lens covers back on to stop the lenses getting scratched and put the binoculars safely back in their case.

When you look at the night sky with binoculars, you will find one big problem. The stars wiggle about all over the place and you'll have to stop looking. It happens because your hands shake slightly. Binoculars magnify the shaking and the stars wobble around crazily.

This can seem very disappointing, but don't worry. There is an answer to the problem. The simplest thing is to hold the binoculars firmly while you are leaning on a wall, fence or anything that is solid.

A better idea is to use a camera tripod and a binocular mount. Both can be bought from a camera shop.

When you fasten your binoculars to a tripod, it is held firmly and the stars are steady. You can move the binoculars around the sky smoothly using the handles on the tripod – you are now using a real astronomical instrument. The bonus is that you can use the tripod with a camera and photograph the sky too (see pages 94 & 95).

Using a Telescope

Buckets of Light

Most people who enjoy astronomy use their eyes and binoculars to look at the sky, but they might also have a telescope. A telescope is really a light bucket; just as a bucket collects water, a telescope collects light.

The first job of a telescope is not to magnify things but to make them brighter and clearer. This is known as the resolution. Magnification comes second.

When you use a telescope, you are collecting light from stars using a big objective lens or mirror. The light is focused, then magnified, by a small lens called an eyepiece. Each eyepiece has a number printed on it, like 40mm, 25mm or 15mm. The numbers show how much each lens magnifies: the lower the number, the higher the magnification. So a 40mm eyepiece is low power, 25mm is medium power and 15mm is high power.

1 Main telescope
2 Finderscope
3 Objective lens
4 Eyepiece
5 Mount
6 Tripod

The golden rule for using eyepieces is to start off with low power (40mm) and change to medium (25mm or 20mm) when the object is in the middle of your view. You can only use a high-power eyepiece (15mm) for very bright objects.

There are two reasons why a high-power eyepiece is not used all the time. On high power we can only see a tiny bit of the sky – it's a small field of view. That means if our star drifts out of sight, it is hard to find again. The other reason is that higher-power eyepieces use thicker lenses and they make the star or planet look dimmer.

Another good rule is to set up your telescope in daylight. When it is set up, you will have a little telescope sitting on top of the big telescope. This is the finderscope. It doesn't magnify much but it has a wide field of view. The idea is to find your star or planet in this little telescope then, when you look through the main telescope, it will be there.

This only works when your finderscope is lined up properly. The best thing is to line it up in daylight by looking at something like a chimney or TV aerial. Adjust the screws on the finderscope and you are ready to go. All you need now is a clear, dark night.

Investigate before buying

If you are thinking of buying a telescope, or someone is going to buy one as a gift, please choose carefully. Before buying, talk to someone who knows about telescopes, perhaps someone who belongs to an astronomical society. When you do buy the telescope, go to a shop that sells only astronomical instruments and listen to their advice.

Quality telescopes are expensive – a good one will cost as least as much as a good digital camera.

Check the mount

Telescopes come on a mount that fits on a tripod. Mounts are complicated things, but any mount you buy must hold the telescope firmly in balance. It should be easy to set up and the tripod must be solid. If you put an expensive telescope on a wobbly tripod, it will be almost useless, before you buy, make sure the telescope has a solid mount and tripod.

Focusing

To see a star or planet or nebula clearly, you must focus the telescope by turning a focusing knob near the eyepiece. Turn the knob until you see a very clear, small image. If the star or planet looks big, round and fuzzy, it is not focused.

Basic Viewing

Stargazing

How to Look at the Sky

Observing the night sky sounds so easy. You just walk outside, look upward and see the stars. But, if you do that, you will probably not see very much. There are ways of making sure that you have a really good view of a starry sky.

Before you go outside, make sure that you are wearing warm clothes. Even in summer, it can get cold at night. In winter, you need lots of layers of clothes, a woolly hat and gloves. Always check with an adult and ask permission before venturing out.

Most of the time, you will probably observe the stars from your yard. Choose the darkest place, somewhere that blocks light from house windows, street lights and security lights. You need a dark spot where you can still see most of the sky.

If you observe away from home, you must be safe, and that means having a parent or a trusted adult with you. If you know you are safe, you will enjoy your stargazing. Actually, it's a good idea to observe with your friends and family, especially when you're doing something like a meteor watch. You might even run star parties in your back yard!

Your eyes need a few minutes to become used to the dark – this is called dark-adaptation, or getting your night vision. At first you will not see many stars. After ten minutes you can see many more, and after 20 minutes your eyes will be properly dark-adapted.

Any bright light will spoil your night vision. You will need a flashlight but make sure that it is a red-light flashlight, which does not spoil your night vision. You can buy a red LED flashlight or just stick some red cellophane over an ordinary flashlight.

Being an astronomer is cool. Actually, it's often cold – but an astronomer must be a cool person too. That means being patient. Clouds often spoil the sky and you must wait and hope for the sky to clear. Some nights you will do more cloud watching than star gazing.

A Stargazer's Menu

On a clear night, check all the main constellations first. Then look at the Moon, which is always impressive through binoculars or a telescope. If there's a planet in view, that comes next. Finally take a look into deep space. With binoculars, look at bright stars to see their colors and observe star clusters too. With a telescope you can search out nebulae among the stars of our Milky Way. You could even look for other galaxies far away.

Observing Deep Space

Stars are so far away that even viewing through binoculars or a telescope they are just points of light. They do look brighter though and you will see true star colors. Star clusters often look best through binoculars where you will see all the stars together. Nebulae and galaxies are dimmer, but to discover objects so far out in deep space is fun.

Observing Planets

Planets are millions of miles/kilometers away, so even viewing through a telescope they look quite small. However, as planets are bright you will see them clearly. Spend time looking at a planet as you will start picking up more detail – you might even see changes from night to night as you observe different parts of the planet.

Photography

Taking Pictures of the Night Sky

It would be great if you could take photographs of what you see. Photography of the night sky is not easy, but it can be done and it's worth trying. You might need an adult to help.

Most photos that show planets, nebulae and galaxies were taken from spacecraft or through huge telescopes.

Today, astronomers don't use ordinary cameras. They use expensive CCD imagers, which are like super digital cameras. Even so, the imagers have to collect light for several minutes or hours, and then the astronomers have to do a lot of computer-processing to get the final photograph.

You may not be able to take close-up photos of planets or deep space, but you can take pictures of the Moon, constellations and shooting stars.

The best place to start is the Moon. You will need a digital camera and a telescope or binoculars and, if you are using binoculars, they must be mounted on a camera tripod. Focus your telescope or binoculars on the Moon and place the camera carefully to the eyepiece. Remember to turn off the camera's flash before you start.

Meteors and Satellites

You can photograph meteors and satellite tracks in the same way that you photograph constellations.
For a satellite, point the camera to where a satellite will pass by.
For meteors, wait for a meteor shower night and point the camera high in the sky. If you are very lucky, you might catch a shooting star!

Star Trails

If you have a camera with a shutter that stays open for several minutes, you can take lovely star trails like the photo on this page. Point the camera at the North Star. Press the button and leave the shutter open for several minutes. As the Earth turns, the stars seem to move and make trails of light around the Pole Star. This shows how quickly the Earth turns around. Any yellow glow will show how much light pollution there is in your sky.

Move the camera gently until you see the Moon on the camera screen. Press the shutter. It may not work first time, but keep trying. You can erase photos that are not good.

If you take pictures of stars and constellations, you will need a camera that lets you keep the shutter open for several seconds. Ask an adult to help you. Set the camera to a 15- or 20-second exposure and put it on a tripod to hold it steady. Point the camera at a constellation and you are ready to start.

If you have a digital SLR camera, you can use the 'B' setting and a cable release. You can take exposures from 10 to 20 seconds. Remember to save your camera's memory by deleting photos that you don't want to keep.

If you don't have an expensive SLR digital camera, you can still achieve good results. Simple compact digital cameras are capable of good night-time picures too. No matter what camera you have make sure that you choose a good location with an unobstructed view of the stars or Moon. It is also best to avoid any stray or obtrusive lighting. If you are taking the shot from your bedroom window, turn off the light. If you are outside, try to avoid standing under a street light. It is always better to use a tripod for this type of picture but if you don't have one use whatever is available to keep your camera steady,. perhaps a window ledge or garden wall. Use the 'long exposure', 'night' or 'fireworks' setting if your camera has one. This should turn off your flash and keep your shutter open for a long time, allowing for optimum exposure. If you camera has a manual mode you can experiment by turning off the flash and increasing and decreasing the length of exposure.

Glossary

Meanings of astronomical words in alphabetical order

Atmosphere

Constellation

Neutron Star

Satellite

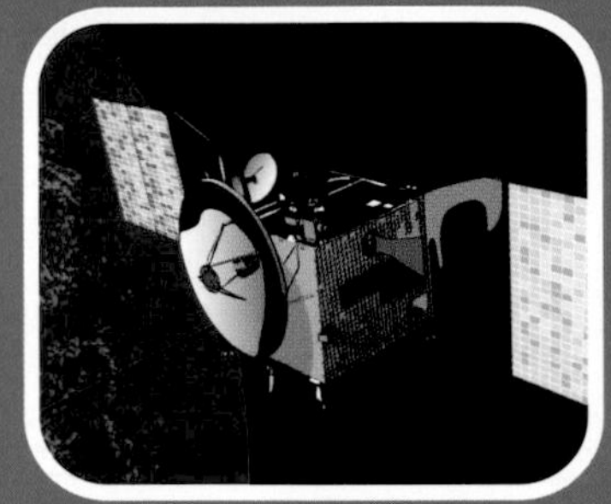

Asteroid A rocky object that can measure a few hundred miles/kilometers across. Most asteroids orbit the Sun between Mars and Jupiter.

Atmosphere A layer of gas around a planet, like the air on Earth. The thickest part of Earth's atmosphere is from the surface to about 9 miles/15 kilometers high.

Black Hole The crushed center of an exploded star with gravity so strong that nothing can escape from it.

Cluster A group of stars or galaxies held together by gravity.

Comet A lump of ice and rock, about 6 miles/10 kilometers across, in orbit around the Sun.

Constellation A group of stars that, seen from Earth, make a pattern in the sky. There are 88 constellations in total.

Galaxy A collection of millions or billions of stars, planets, gas and dust held together by gravity. Our galaxy is the Milky Way.

Gravity A natural force that pulls objects together.

Kuiper Belt An area far out in the Solar System orbited by dwarf planets such as Pluto.

Light Year The distance traveled by light in one year. One light year is equal to 5,903,000,000,000 miles/ 9,500,000,000,000 kilometers.

Meteor A piece of space dust, that is no bigger than a grain of sand, that burns up in the atmosphere. It is also known as a 'shooting star'.

Meteorite A piece of space rock which hits the surface of a planet or moon. They measure from a few inches/centimeters to several miles/meters across.

Nebula A cloud of gas and dust between the stars.

Neutron Star The crushed center of an exploded star, that weighs from one and a half to three times more than the Sun.

Oort Cloud Millions of comets in orbit around the Sun at the edge of the Solar System.

Orbit The track of an object around another object, like a planet around the Sun, or a satellite or moon around a planet.

Pulsar A spinning neutron star that sends beams of energy into space.

Red Giant A huge star near the end of its life, examples are: Betelgeuse in Orion and Aldebaran in Taurus.

Revolution The movement of a planet in orbit around the Sun.

Rotation The spin of a planet or star on its axis.

Satellite An object that orbits a planet. Earth has many man-made satellites and one natural satellite, the Moon.

Supernova The explosion of a supergiant star, leaving a pulsar or black hole.

White Dwarf The squashed, heavy middle of a dead star.